Small Stock Management

J. Oberholster

Small Stock Management

Terminology

Management

Selection

Nutrition

Health

Economy

Small Stock Management

Author
Johan Oberholster

Publisher
Kejafa Knowledge Works
Noordheuwel
Krugersdorp
011 025 4388
kejafa@mweb.co.za

Printer
Craft Print International Ltd.
Singapore

Photos
Johan Oberholster

Cover design
Christina Harman
blackpepperdesign

First Edition
September 2010

ISBN
978-0-620-47896-0

Introduction

Sheep and goats are of ancient origin in Southern Africa and date back more than 2000 years. Topography and soils influence animal production directly. Infectious diseases and internal and external parasites are severe constraints in sheep and goat production and they are endemic to large regions in Southern Africa. In financial terms it is calculated that reproduction is twenty times more important than carcass qualities, and ten times more important that those characteristics associated with growth and feed conversion. The reproduction tempo further determines the intensity of selection that can be applied and consequently the genetic advances that can be made. As reproduction has a low heredity (10%), (*Maree and Casey, 1993*), it is the reproduction tempo that determines the financial and economic success. By today's standards of world population and settlement density, it is no longer possible of making way with ample compensation. New agricultural ways had to be found, supported by technical and veterinarian developments. What once was a visionary dream - improving food production enhanced by pharmaceutical influence - has become a real possibility. Obviously this led to, and will lead to, a search for answers. Recommendations formulated in many ways have already confronted one another, not only in past years. Finding correct solutions depend on how the human community reacts to such a challenge, and organizes its production, and agricultural process. The lack of significant improvement regarding the ever-increasing demand for mutton meat has been of concern for some time. Increased economic pressure calls for intensifying the production of ewes. For increased flock fertility, adaptability and reproduction efficiency are major factors. From an economical point of view, a high fertility with regular lambing under good breeding conditions and speedy weight increase is seen as a very important criterion in Small Stock production. The associated problems of low live mass weight are known: weak lambs, poor survival chances and an increased mortality.

About the author

Johan Oberholster started his career in 1988 as a Small Stock Research Technician in Government Service Namibia. He first did a National Diploma in Agriculture, and then B.Tech degree in Animal Production at the Pretoria Technicon. During these 19 years he worked in various positions on four government Research Stations. He gained extnsive experience in practical farming and farm management since he worked in management positions on all of these farms. He worked with large stock, small stock (namely Dorper, Damara and Karakul sheep, and Boergoat stud), Crop Science and dairy environments. Duties includes, amongst other things, grazing management, maintenance of infra-structure, record keeping, artificial insemination, disease and predator control, drawing up and control over small stock budget and personnel management within the section. He spent the past 14 years on Neudamm Agricultural College in Namibia. The past 10 years he acted as small-stock ring official or judging animals at the Windhoek Show. He is also involved in practical training of students from University of Namibia, the Polytechnic of Namibia, the training of extension technicians in small stock management at Katima Mulilo, Ongwediva, Opuwo and Rundu, those employed in the Namibia Northern Communal Areas, as well as pupils from Windhoek Technical High School. He also gave t echnical assistance with regard to small stock management, and wrote a manual on this topic in line with needs and specifications for the curriculum of Agricultural learners at Higher Education.

Dedication

I'd like to dedicate this book to my Parents and my Creator and Lord

Gen 24:35 And the LORD hath blessed my master greatly; and he is become great: and he hath given him flocks, and herds, and silver, and gold, and menservants, and maidservants, and camels, and asses.

Gen 24:35 En die HERE het my heer baie geseën, sodat hy groot geword het, en aan hom kleinvee en beeste, silwer en goud, slawe en slavinne en kamele en esels gegee.

CONTENTS

Small Stock Management

J. Oberholster

CHAPTER 1: INTRODUCTION

Since the earliest days sheep and goats have played an important role in the livestock industry in Namibia. As seen from Fig. 1 and 2, sheep constitute approximately 54.3 % of the total numbers of small stock or 33.2 % of all livestock, while goats constitute approximately 45.7 % of small stock or 28 % of all livestock in Namibia.

TABLE 1.1 SMALL STOCK NUMBERS IN NAMIBIA 2002.

TYPE	SOUTH	CENTRAL	NORTH	TOTAL
DORPER	1 500 970	418 132	12 019	1 931 121
KARAKUL	207 989	12 969	1 874	222 832
ANGORA	4 544	0	0	4544
BOER GOAT	396 681	339714	137 355	973 750
OTHER GOATS SHEEP	295 168	312354	224 672	832 194
TOTAL	**15 914 082**	**1 083 169**	**375 920**	**17 373 171**

TABLE 1.2 LIVESTOCK NUMBERS IN NAMIBIA 2002.

TYPE	SOUTH	CENTRAL	NORTH	TOTAL
SHEEP	991 316	691 535	236 258	1 919 109
GOATS	229 902	391 634	1 028 993	1 650 529
CATTLE	99 352	452 020	1 580 019	2 131 391
TOTAL	**1 320 570**	**1 535 189**	**2 845 270**	**5 701 029**

TABLE 1.3 SMALL STOCK STUD BREEDERS NUMBERS IN S.A.

(SA Stud Book Association – Small Stock Breed Count 01/07/2006)

TYPE					
SHEEP AND GOATS		**Breeders**	**Male**	**Female**	**Total**
AFRINO	AFO	3	56	166	222
ANGORA	AGG	19	1190	1245	2435
ALPACA	ALP	33	148	365	513
BRITISH ALPINE	BAG	*	21	198	219
BOERBOK	BBG	22	2763	3743	6506
BORDER LEICESTER	BLS	-	-	-	-
CORRIEDALE	COR	2	85	175	260
DAMARA	DAM	13	114	334	448
DOHNE MERINO	DMS	93	204	256	460
DORMER	DOM	59	2633	7597	10230
DORPER + WIT DORPER	DOP	177	5896	62687	68583
DORSET	DOR	1	22	53	75
BUITELANDSE DORPER	DPS	1	107	127	234
GORNO ALTAI	GAG	-	-	-	-
HAMPSHIRE	HSS	4	66	144	210
ILE DE FRANCE	IDF	53	994	3467	4461
KARAKUL	KAR	2	2	783	785
KALAHARI REDS	KRG	5	133	440	573
LETELLE	LET	3	95	74	169
MERINO	MER	17	2510	2611	5121
MERINOLANDSKAAP	MLS	33	1245	4613	5858
MEATMASTER	MMS	3	98	428	526
PEDI	PED	8	92	212	304
ROMANOV	ROM	2	67	111	178
SA VLEIS MERINO	SAM	1	0	0	0
SAFFER BOKKE	SFG	1	84	94	178
SAANEN (*)	SMG	16	122	474	596
MELKSKAPE	SMS	1	9	51	60

SAVANNA GOATS	SSG	3	2	24	26
SUFFOLK	SUF	11	157	724	881
TOGGENBERGER	TOG	*	17	189	206
VANDOR	VAN	-	-	-	-
VAN ROOY	VRS	-	-	-	-
TOTAAL		**588**	**18932**	**91385**	**152936**

* *All Milch Goat Breeders = 16 (Number opposite Saanen).*

1.2 INCORPORATION OF SMALL STOCK IN A CROPPING INTERPRISE

The incorporation of livestock enterprise has a stabilising effect on crop farming.

Marginal lands are used more profitably. The animal factor usually ensures a better cash flow.

1.2.1 ADVANTAGES OF INCORPORATING SHEEP RATHER THAT OTHER ANIMALS

- Turnover is faster and the cash flow is better with sheep than with cattle.
- Cash outlay for creating facilities for sheep is smaller than for cattle.
- Sheep farming can start on a smaller scale than cattle farming.
- Sheep farming is not as management intensive as a dairy enterprise.
- Sheep make better use of harvested land since they pick up wasted grain.

1.2.2 MANAGEMENT SKILLS

- The owner and his subordinates must have the necessary knowledge of sheep management.
- Sheep moved from an area must be immunised against pulpy kidney and the diseases prevailing in the new environment.
- Dosing against internal parasites is also essential. If there are vast differences between the old en new environment, it is desirable to dose new small stock with the rumen juice of local sheep.

1.2.3 SECURITY

Theft, dogs and vermin can lead to substantial losses, which is why security facilities and practices must receive attention.

1.3 SHEEP TERMS AND DEFINITIONS

1.3.1 EWES MATED

This is the number of ewes in the flock set aside for mating. If 500 ewes were available for A.I., hand service or group mating, but only 400 came into season, then for the purpose of calculation lambing or weaning percentage, it is still taken that 500 ewe ewes were mated.

1.3.2 LAMBING PERCENTAGE

The number of lambs born expressed as a percentage of the number of ewes mated, e.g. 500 ewes mated and 600 lambs born.
Lambing % = 600/500 x 100/1 = 120 %

1.3.3 WEANING PERSENTAGE

The number of lambs weaned expressed as a percentage of the number of ewes mated, e.g. 500 ewes mated and 600 lambs born. But 60 lambs still – born or died prior to weaning i.e. 540 weaned
Weaning % = 540/500 x 100/1 = 108 %

1.3.4 FECUNDITY

The numbers of lambs born per ewe lambed, e.g. 500 ewes mated, only 400 lambed, but 600 lambs born.
Fecundity = 600/400 = 1.5

1.3.5 CONCEPTION PERCENTAGE

This is the number of ewes lambed expressed as a percentage of the number of ewes mated, e.g. 500 ewes mated, but 400 lambed and aborted.
Conception percentage = 400/500 x 100/1 = 80 %

1.3.6 FEED CONVERSION

This is the amount of feed consumed by the sheep for 1 kg live-mass gain, e.g. a lamb over a certain period consumes 20 kg feed for a live - mass change from 30 to 35 kg.
Feed conversion = 20/ (35-30) = 3.1 kg

1.3.7 AVERAGE DAILY GAIN

This is the animal's daily gain over a certain period, e.g. from birth to wean. If a lamb had a mass of 4 kg at birth and after 100 days at weaning it had a mass of 25 kg, its ADG would be (25-4)/100days = 0.21 kg or 210 g/day.

1.3.8 LAMB MORTALITY

Number of lambs that died between birth and wean expressed as a percentage of the total number of lambs born, e.g. Mortality % = 30/600 x 100/1 = 5 %

1.4. MATING METHODS

Do not make ewe groups too big. Flock splitting may occur within a camp and ewe groups may be left without a ram, resulting in a low conception rate.

It is also good practice to put small groups in small camps for mating. Make sure that the mating group is herded together in the early morning and late afternoon.

1.4.1 GROUP MATING

Use 2/3 % rams with 100/150 ewes for six-week period. When teaser rams are used to synchronize the ewes, the mating should be four weeks.

1.4.2 HAND MATING

Hand mating is very labour intensive and can be very time consuming. Rams must be trained. Proper handling facilities with sufficient holding pens are essential. On the other hand mating can be done with less rams, this will result in more genetic advance. When selecting ewes for hand mating by making use of teaser rams, it should be done in the handlings pens, under direct supervision. Immediately after that, the cycling ewes should be mated.

For hand mating, A.I. and individual mating, a mating register must be kept.

Particulars that are entered into the register are:

- Date mating starts
- Date mating ends
- Number of ewes mated
- Ear numbers of ewes and rams used
- Numbers of ewes lambs
- Lambing percentage

Such a register is important in stud breeding.

Disadvantages of this method are:

- Labour intensive
- Time consuming
- Pasture around the farmstead trampled
- Rams trained

Advantages of this method are:

- Saving on ram cost
- Large group of good rams can be obtained
- Reliable record can be kept
- All ewes on heat can be serviced
- Rams with a low libido and sterile ewes and rams can be identified
- Rams can be fed special rations and kept active during mating time

1.4.3 INDIVIDUAL MATING

Individual mating is mostly use by stud and commercial farmers who want to know the genotype of their lambs. The ewes are synchronized with teasers. A fertile ram is joined to 50/100 ewes in a small camp. It is important to monitor the ram performance with marker block. An infertile ram can be the reason why a lot of ewes do not conceive. Bring the groups in daily to clean the blocks. The ram must be fed high protein.

1.4.4 ARTIFICIAL INSEMINATION

Definition: Artificial insemination is the fertilization of the female ovum (egg cell) by male sperm cells without sexual intercourse (coitus) having taken place.

1.4.5 Advantages

- Outstanding male animals can be used to their full potential.
- Because no direct contact takes place, venereal diseases can be controlled.
- An outstanding male animal that is old or injured can still produce progeny through A.I.
- Instead of a large number of rams, a few good once are needed, which reduce the ram cost per lamb.
- Young rams can be tested at an earlier stage because their progeny are much more numerous.

1.4.6 Disadvantages

- If the ram is not good enough, it is possible that he can do more harm than good with A.I. The poor characteristics are then spread much more widely through the flock.
- A.I. requires a lot of time and labour.
- Negligence during the application of A.I. can result in a large scale spreading of venereal diseases.
- If conception rate is disappointingly low, lambing percentage will be below 60 % and the farmer will suffer a financial loss.
- Knowledge and experience is necessary for A.I. to succeed. A.I. causes a concentration of animals around the farmstead.

1.4.7 MATING SEASON

Most of the small stock breeds are seasonally poly-estrus (they undergo several cycles per season), and display increased sexual activity during autumn. A high lambing percentage can be expected if ewes are mated during the period February to June. Unfortunately the lambs are born in the poor period of the year, which makes it difficult to raise them.

1.4.8 PERIOD OF MATING

The period over which the rams are left with the ewes, should be long enough to allow every ewe the chance to mate. The mating season should not be shorter than 42 days (two estrus cycles).

1.5. RAM MANAGEMENT

Rams are usually scarce and expensive and must be cared for in such manner that maximum utilisation is possible. They must be in a good condition throughout the year so that they can always have a good service capability. Rams must never be too fat, but still in a good condition. Six weeks before mating, supplementary feeding can be given to rams.

1.5.1 FOOT CARE

- Feet must be checked and trimmed every six weeks.
- Trimming must be done two weeks before mating.
- Regular trimming will prevent foot rot development

1.5.2 RAM FITNESS

Ram fitness must be maintained throughout the year and must be intensified two months before mating.

1.5.3 RAM HEALTH

- Inoculations must be kept up to date.
- The last blue tong inoculation must be administered two months before mating.
- A monthly physical screening of the rams must be carried out.
- Internal parasite infections must be monitored by means of dung sample collection for egg counts.

1.5.4 SHEARING AND DIPPING

- These activities must be performed at least two months before mating

- Rams should be shorn every six months, an injection against ectoparasites are preferable to dipping during winter.

1.5.5 FERTILITY

- Fertility test on a ram can be a harrowing experience. If done two months before mating it allows sufficient time for a retest, treatment and / or corrective feeding.
- Genital organs should be palpated and checked two weeks before mating.

1.5.6 CONCENTRATE FEED

Feeding with concentrates, preferably protein should commence two months before mating, this will result in excellent semen quality.

1.5.7 REPLACEMENT RAMS

Replacement rams should be bought well before the breeding season in order to give the rams time to acclimatize to their new surrounding.

1.5.8 TESTES SIZE

Why is testes size important?

- Each gram of testicular tissue produces 20 million sperms per day.
- A ram with a scrotal circumference of less than 24 cm cannot be successfully joined to 50 ewes and should be culled.
- A ram with a scrotal circumference of 30 cm and more can be successfully joined to 100 ewes, provided that the ram is in good health and has a high serving capacity.
- Testes size and fertility is highly correlated, and is reflected in the fertility and sexual maturity of the ewe lamb progeny.

1.5.9 DISEASES

Small stock should as far as possible be vaccinated eight weeks before mating. The following are, however diseases that should receive attention when farming with small stock.

- Brucellosis
- Rift Valley Fever
- Wesselsbron
- Blue Tongue
- Corynebacterium
- Pasteurellosis
- Rabies
- Pulpy Kidney

1.5.10 PARASITES

Parasites are organisms that live at the expense of their host, causing injury to it and given it nothing useful in return. The injury to the host may be only slightly or so severe as to be fatal, with a correspondingly heavy economic loss to the owner.

1.5.11 ENDOPARASITES

This term describes those parasites that live in the internal organs of the host animal, and the extent of damage they can cause depends on the severity of the parasite invasion. To make it possible to farm profitably with domestic livestock, the following preventative and corrective measures should be taken:

- Feed a balanced ration to ensure good nutrition.
- Frequently remove all bedding and manure from contaminated yards and buildings.
- Raise food and water troughs above the ground to prevent contamination with manure.
- Separate obviously infected animals from the rest of the herd.
- Do not overstock pastures and yards.
- Do not graze young stock on pastures previously used by parasite infected animals.

1.5.12 ECTOPARASITES

This term describes those parasites that live on the surface of the host's body and includes ticks, lice, mites and flies. They feed on the blood of their host and are capable of taking in many times their weight in blood.

The dipping of small stock is an essential practice, which is sometimes neglected. The best time to dip the animals is during March or just after the sheep have been shorn. Under certain circumstances it may also be necessary to dip animals during October.

1.5.13 HOW TO DETERMINE THE SERVING CAPACITY OF RAMS

- The serving capacity of a ram must be determined when he has adapted to the farm environment and before he is put with the ewes.
- Only fertile rams that have proven themselves should be used for breeding.
- Once a ram has shown interest in a ewe, that interest will remain, providing that the ram is in good health.
- The younger a ram when he makes sexual contact with a ewe, the greater chances are that he will develop an acceptable serving capacity.
- The younger a ram when he makes sexual contact with a ewe, the lower the chances are that the ram will become homosexual.

1.6 EWE MANAGEMENT

The ewe-flock on any farm can divided into five groups in terms of their productions. A farmer who follows a system of more than one mating season, have four or all of these groups on his farm at any given stage. From the grazing viewpoint it is essential to keep the numbers of flocks as low as possible, unless there are enough camps.

1.6.1 FEMALE WEANERS

The young female wearers are the future producers of the farm and may be neglected. The faster they grow, the sooner they can start reproducing without detrimental effects. It is thus essential for the farmer to bring his young ewes in reproduction stage as soon as possible. For this reason it is essential for young ewes to be fed with protein lick during dry months, and even with supplementary feeds if the conditions of the animals are poor. Young animals must receive special attention regarding their effective control of endo- and ectoparasites.

1.6.2 DRY EWES (NON - PRODUCTION EWES)

These are ewes whose lambs have been slaughtered or weaned, until they are mated again. The nutrition requirement of the group is the lowest and such ewes usually also require the least attention. It must however be borne in mind that these ewes must be in a good condition once the mating season starts.

Important vaccinations must be given during this period, which might cause abortions such as Bluetongue. Other important vaccinations to stop abortions such as Enzootic Abortion must be administered one month before the mating season starts.

1.6.3 EWES THAT ARE MATED

It is desirable to separate ewes that must be mated, from other ewes. This ensures a smaller flock and the ram can find ewes in heat easily. Ewes should also receive good feeding at this stage. The camps in which mating take place, should be rather small and not too dense or bushy, so that the rams will have a better chance of tracing the ewes in heat. Phosphate lick must be available and protein lick during the dry season.

1.6.4 PREGNANT EWES

The gestation period of small stock is approximately 148 to 150 days. The nutritional needs of ewes in lamb are much higher than those of dry ewes. Supplementary feed must be considered during dry seasons, particularly in view of better milk production of the ewes after lambing. Three weeks before lambing season starts, supplementary feed like maize with high protein value can be used to help milk production increase in ewes.

Ewes which are poorly fed during this period, will have weak lambs at birth and this can lead to high mortalities and retarded growth. Ewes in lamb must be handled carefully or preferable be left alone altogether. Late pregnant ewes should rather not be vaccinated or dosed, because it can cause abortions.

1.6.5 EWES WITH LAMBS

The lambing season is a critical period in any farming operation. Ewes should lamb in small camps of 10 to 15 hectare close to the farmstead. These camps must be easily accessible and fairly free of shrubs so that ewes that have given birth can easily be located. It is also not desirable to allocate more than approximately 50 ewes to a camp.

If more ewes are placed in a camp, it is preferable to move the ewes, which have not lambed, to another camp on a weekly base. The nutritional requirements of ewes with lambs are very high and therefore good grazing must be available. If grazing conditions are poor, supplements must be fed.

1.6.6 INCREASED LAMBING RATE

In general the intensive lambing frequency, where the same ewe lambs more than once a year, it is not widely applied in Southern Africa. The available breeds in Southern Africa with a prolonged breeding season, such as e.g. the Merino, S.A. Mutton Merino and Dorper make it possible to have three lambing seasons over two years without making use of any hormonal treatments. To achieve this increased lambing frequency, it is preferable not to allow mating on fixed 8 – monthly intervals, but rather to incorporate it on a seasonal basis.

1.6.7 REPLACEMENT EWES

A higher replacement rate has a direct influence on the age composition of the flock and genetic progress with selection. With a ewe replacement of 20 to 25 % per annum, strict selection can take place and breeding ewes can be eliminated at age 5 or 6 years. The culling age will be determined by the percentage and the relative economic value of surplus ewes of different ages. A ewe replacement of 20% is generally applied, which ensures that one can select more strictly among young ewes. Fewer replacement ewes need to be carried over to two-tooth age, which thus ensures the highest percentage breeding ewes.

1.7 RED MEAT PRODUCTS

1.7.1 INTRODUCTION

Red meat is named according to their source: beef is typically from cattle over a year of age. Veal is from calves 5 month of age or younger, pork is from swine, mutton from mature sheep, lamb is from young sheep less then one year, chevon from goats, but it is more commonly called goat meat.

The Meatmaster breed.

1.7.2 MUTTON PRODUCTION

The mutton industry basically consists of two important aspects, (1) the number and (2) quality of the carcasses that are produced. The number or surplus stock that can be produced will mainly be determined by the following factors:

- the composition of the flock
- the age of ewes at the first and final lambing
- the number of lambs weaned from mated ewes
- lamb mortalities
- increased lambing rate

Flock compositions refer to the number of rams, mature ewes, and young ewes for replacement, lambs and surplus animals in the flock. The mutton production potential of the flock depends on the number of ewes that is available for breeding. It is also clear that the higher the fertility in the flock, the lower the number of ewes that are required for breeding. Those ewes will be able to produce the same number of slaughter lambs, or even more. Fertility has the single most important influence on any mutton production system.

The meat production potential of a flock can be increased by mating young ewes, so that they can produce their first off-spring at the age of one year. With regard to the effect of early breeding on the later reproduction capacity of ewes, those ewes that gave birth for the first time as yearlings showed a greater life-span production potential then ewe-lambs that first lambed at the age of two years. Ewes that showed follicular activity, whether or not they were mated as lambs, had a greater life production capacity than ewe-lambs that showed no signs of puberty during the first year.

One way of increasing the litter size, is by synchronizing the estrus and by stimulating the rate of ovulation by way of exogenous therapy. However problems with regard to cost, reduction of conception capacity and the unpredictability of the litter size restrict the practical application of the method. The quickest method of achieving genetic improvement is by crossbreeding the indigenous breeds with the highly fertile breeds.

At present it seems as though the only viable method to improve fertility is through the selection of the most fertile animals within a breed for breeding. Research work done at Stock Breeding Research Organization in Edinburgh, has shown that fertility in sheep can be determined not only by the number of lambs produced by mature ewes, but also by measuring the rate of increase in the size of the testicles of young rams.

Investigations in most of the small stock producing countries indicated that between 15 and 20 % of the lambs born, died between birth and wean age. Most of these lambs are either stillborn or died within 3 day to five days after birth. Inheritable infections may be responsible for losses. The direct economic loss in lambs is thus considerable and can be expressed in monetary terms. These losses include the following:

- supplement feed to pregnant ewe lost
- ram costs per lamb increases for the remaining lambs
- a decline in breeding progress due to the loss of future breeding material.

1.7.3 SLAUGHTER LAMB PRODUCERS

The distance from the nearest market has a considerable influence on the profitability of slaughter lamb production. If slaughter lambs have to be transported over too long distances, it is not only results in additional cost, but there is more weight-loss during such trips. The quality of the carcasses is also affected, especially as a result of injuries and bruises the animals sustain during transport.

In order to change and maintain a specific flock composition, the farmer has to establish and stick to a predetermined marketing system. To prevent over stocking, all weaned lambs must be marketed as soon as possible, except young ewes required for replacement. The longer the ewes are retained for breeding purposes, the smaller the replacement percentage will be and higher the number of lambs for marketing.

1.7.4 MAKING EFFECTIVE MANAGEMENT DECISIONS

Effective management of livestock operations implies that available resources are used to maximize net profit while the same resources are conserved or improved. Available resources include fixed resources (land, capital and management) and renewable biological resources (animals and plants). Most livestock producers manage their operations with plans to make profit.

Simply stated, the profitability formula is:

profit or loss = (production x price) - cost

The formula can be expanded to make management decisions more focused:

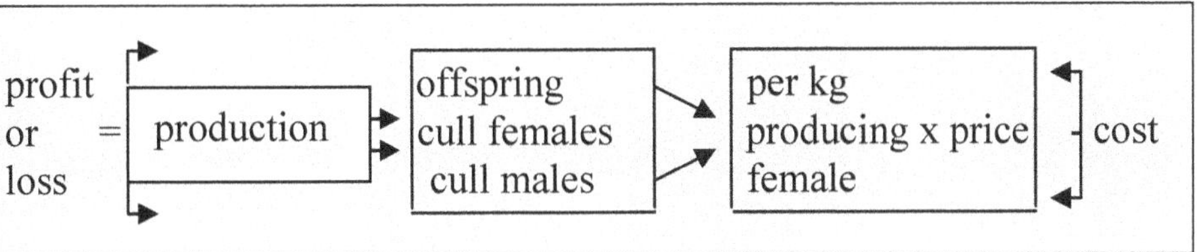

Cost – include feed, workers, veterinary, repairs, fuel and interest, obviously profit occurs when output value exceeds input costs and loss occurs when input cost exceed output value. Price – the amount received per kg, per animal, price is influenced primarily by supply and demand. Production / output – output is usually kg / numbers sold.

An effective manager, whenever involved in a one –person operation or a complex organization structure, needs to:

- be profit-orientated
- identify objectives of the business and establish both short-term and long-term goals to achieve those objectives
- keep abreast of the current knowledge related to the operation
- know how to use time effectively
- attend to the physical emotional and financial needs of those employed in the operation
- incorporated incentive programs to motivate employees to perform at their full capacity each day
- have honest business dealings
- effectively communicate responsibilities to all employees and make employees feel they are part of the operation
- know what needs to be done and at what time
- be a self-starter
- set priorities and allocate resources accordingly
- remove or alleviate high risks
- set a good example for others to follow

1.7.5 FACTORS DETERMINING THE VALUE OF THE CARCASS

Sheep are classed according to age, sex, and grade. Age classes are lamb, yearling and sheep. Lambs are further divided by age as hothouse lambs, spring lambs, feeder lambs, breeding sheep and shearer lambs. The sex classes are ram, ewe and wether. Hothouse lambs - are under 3 months of age and usually weigh less than 27 kg, they are also for specialty market between Christmas and Easter. Spring lambs – are those 3 to 7 month of age, they are finished at 31 – 40 kg. Spring lambs are born in the fall and marketed in the spring and early summer. Feeder lambs – are those of 7 to 12 month of age, they finish at 39-58 kg; they are usually milk and grass fed.

The grading of carcasses is done by experienced and well trained graders at certain abattoirs according to general standard.

TABLE 1.4 ROLL MARKS ON BEEF, LAMB, SHEEP AND GOAT

Roll mark	Color	Age	Tenderness
AAA	Blue	A	Tender
AAA			
BBB	Green	B	Less Tender
BBB			
CCC	Red	C	Least Tender
CCC			

All goat carcasses are roll marked in orange.

1.7.6 THE FOLLOWING SERVE AS INDICATIONS OF HOGGETING IN SMALL STOCK

There are variations occur within a breed, nutrition as well as breed can also play a roll.

2 teeth - 15/18 months of age
4 teeth - 24 month of age
6 teeth – 30 month of age
8 teeth – 36 month of age or older.

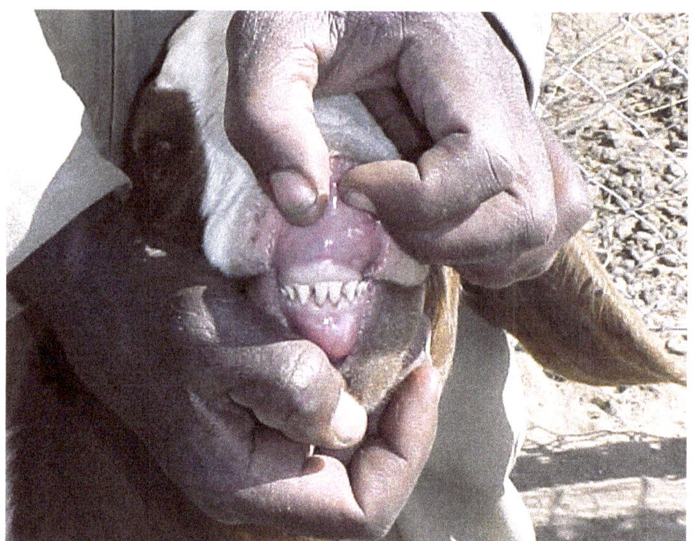

Fig. 1.1 Immature goat lamb 12 month of age

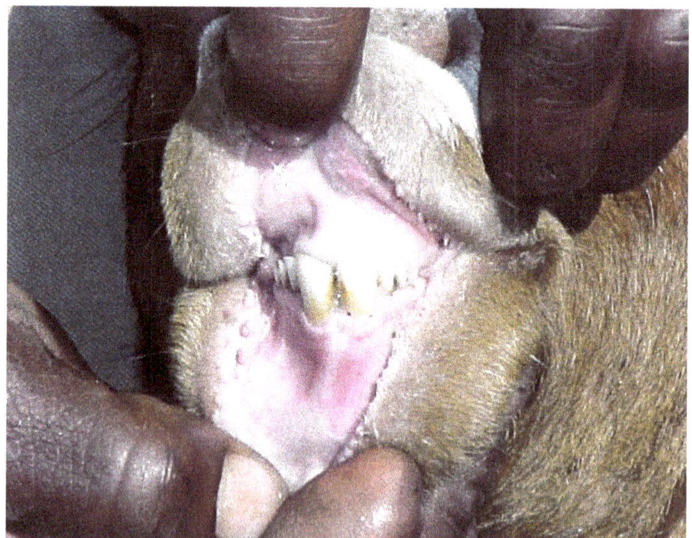

Fig. 1.2 Two teeth – 15-18 month of age

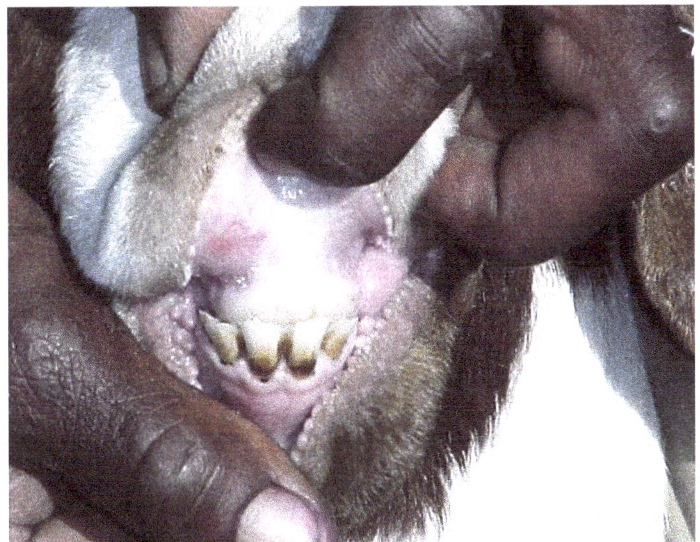

Fig. 1.3 Four teeth – 24 month of age

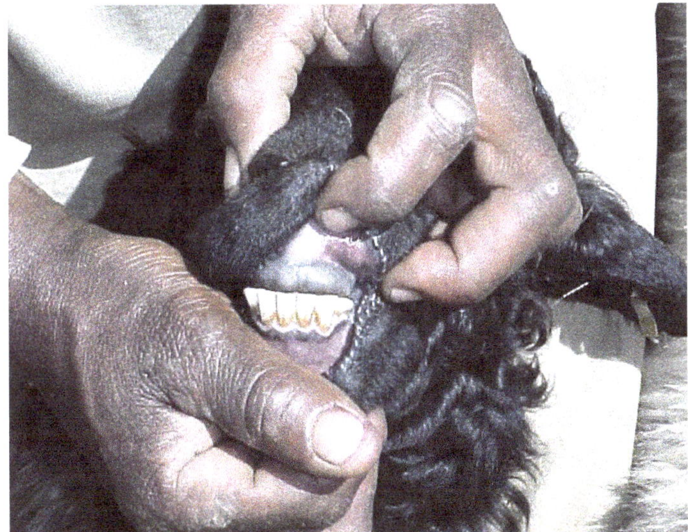

Fig. 1.4 Six teeth – 30 month of age

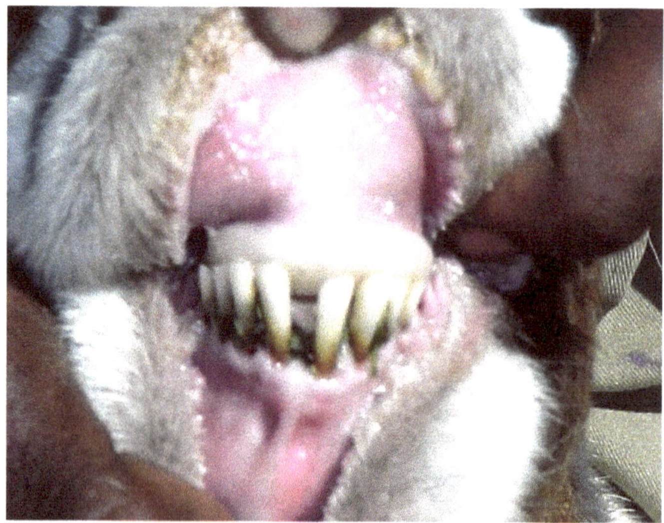

Fig. 1.5 Eight teeth – 36 month of age or older

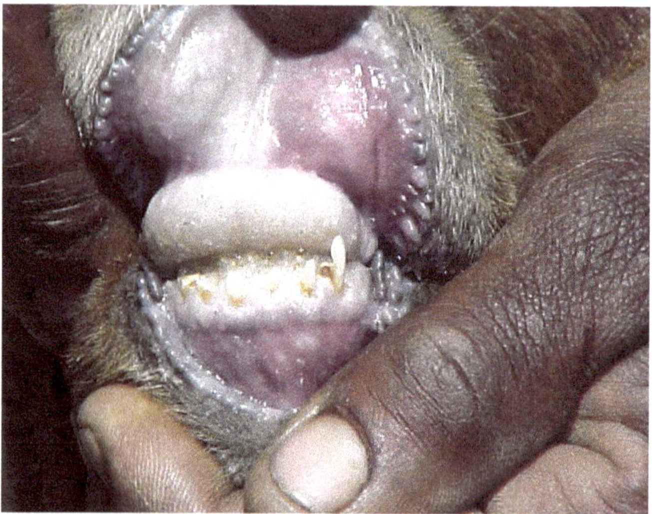

Fig. 1.6 Six years of age or older

1.7.6 FAT

The subcutaneous fat or visible fat on a carcass is an exceptionally reliable measure of the fat content as well as the meat content of the carcass. The distribution of fat over the whole body is also taken into consideration.

TABLE 1.5 FAT CODES SERVE AS GUIDELINES FOR THE GRADING OF CARCASSES

Fatness code	Description	Subcutaneous fat	
		Cattle	Sheep
0	No fat	-	-
1	Emaciated	<1mm	<1mm
2	Lean	1.1 – 3 mm	1.0 – 4 mm
3	Medium	3.1 – 5 mm	4.1 – 7 mm
4	Fat	5.1 – 7 mm	7.1 – 9 mm
5	Over fat	7.1 – 10 mm	9.1 – 11 mm
6	Extremely over fat	> 10 mm	> 11 mm

1.7.7 DAMAGE

Damage is indicated by three classes (1 to 3) according to visual observation of the damage to the bone – to –fat –muscle ratio. These classes are:

Class 1 - slight

Class 2 - medium

Class 3 - severe

The area where the damage occur is also indicated, via B (round), L (loins) and F (forequarters)

1.7.8 SEX

The sex of animals is only indicated with the code M/D in regard of bull and ram carcasses in the C-age grades.

1.8 MILK AND MILK PRODUCTS FROM SHEEP AND GOATS

Commercial milk production is important in many southern European central and eastern European, and Middle East countries. Sheep milk may contribute directly to the human diet in many countries of northern Africa and the Middle East. In most of the cases where sheep are being milked, the ewe still provides milk for lamb growth for the first 4-6 weeks of lactation. Ewes with twins produce 30-50% more milk than ewes with a single lamb. The two most critical nutrients to support milk production are energy and protein, due to the high fat content of sheep milk (4-7%) and high level of milk sugars of sheep's milk, energy is the major limiting nutrient affecting milk synthesis. Both Ca and P are notably at high levels in milk and probably the most critical minerals to the lactating ewe.

Dairy products from goat and sheep are goat/sheep milk, cheese, butter and yogurt.

Milk goat's numbers are increasing rapidly in the USA, but goat milk remains a specialty product and does not currently rival dairy cow milk. The average herd is small, but several large goat dairies are located near metropolitan areas and modern equipment adapted for goat is readily availably. Similar to Angora being for mohair, milk goats are selected for milk yield with less emphasis placed on other characteristics. Reproductive rate in milk goats is higher than in Angora goat, but probably less than in meat. Fat content of goat milk is reported at 4,2 %.

1.9 THREE MATING SEASONS OVER TWO-YEAR PERIOD FOR SMALL STOCK

1.9.1 JANUARY

1.9.2 Rams

It is usually easier to accurately determine fertility in rams than in ewes, and it is also considered to be more hereditary than in ewes. The ram makes the biggest contribution towards general fertility in the flock, and, if ignored, have far reaching consequences for the enterprise. Therefore, the ram is usually suspect number one when fertility problems are experienced in the flock.

Management practices in rams.

Before commencement of the breeding season, all rams should be subject to fertility testing and clinical examination. Rams with deviations should not be used. Seed should be tapped and tested for live sperm, maneuverability, density, colour and abnormality.

The condition of the rams should be maintained throughout the year while "flush feeding" should commence about one month before the mating season starts. Maize can be given as creep feeding, 100g/h/d. – (week – 1, week –2, rest of time 150g/h/d).

Overweight rams are not fit and this also has in influence on their fertility.

New rams should be bought from breeders who do performance testing, and a certificate that guarantees fertility must be obtained. Rams with a birth mass lower than the average birth mass of the flock, but with an above-average growth, should be bought.

Continuous selection for different stages of growth inevitably leads to an increase in birth mass, and must therefore be monitored on a annual basis.

Temperature (high/low) causes temporary infertility, because ripening sperm dies. It is recommended that rams are being sheared during summer and placed amongst ewes during night time. Farmers must also keep in mind that rams with long testicles produces better semen that rams with short testicles. Rams with a greater testicle circumference, have a higher libido.

Deworm rams, and treat against external parasites. Rams must be vaccinated with Brucella Rev 1 at weaning age.

Vaccinate rams one month prior to mating.
On the day when mating starts, vaccinate rams with Multimin – E to prevent shortages of minerals and vitamins, and to increase fertility.

Decide which rams to use on which group of ewes, as well as the number of rams per ewe group. If rams are going to be used in AI, then those rams must be familiarized to the area in which AI is going to take place, as well as to people, two to three weeks before AI commences.

Young or inexperienced rams can be placed amongst cull ewes for two to three weeks to teach them mating proficiency. A few old rams amongst the young rams would also help. A greater percentage of young and inexperienced rams are required in order to achieve higher conception and lambing percentages.

The recommendation is for 4 % young rams. There is a correlation between testes size and sperm production. Generally speaking, the testes of indigenous breeds like the Dorper are better developed than the testes of exotic species.

1.9.3 EWES

Vaccinate ewes one month prior to mating – Enzovax (every two years), Multimin – E to prevent shortages of minerals and vitamins, and to increase fertility.
Deworm ewes, and treat against external parasites.
Ewes must be vaccinated with Brucella Rev 1 at weaning age.

Fertility is a good criteria to use in order to determine whether animals are adapted to their environment. As fertility increases, total weaning mass increases as well, while production cost per kg decreases. The value of fertility lies in the fact that more animals are available for marketing while selection intensity can be increased because the pool, from which animals can be selected, is greater.

1.9.4 FERTILITY

Although fertility is not highly transferable, it does not mean that one should not attempt to select for this trait. It is better to remove those animals that are not fertile from the flock. Selection for fertility in rams is essential because of the indirect response which male fertility has on the fertility of female animals.

1.9.5 REPRODUCTION LIFETIME

Highly productive ewes with a long reproductive lifetime offer the ideal opportunity to select for progeny with above average production characteristics and which can then serve as replacement ewes. It is also essential that, even in the case of cross-breeding, the mother line should be selected from animals with a long reproductive lifetime. An efficiency index for each ewe, together with an evaluation of the condition of her teeth and general body condition, can be used to cull ewes.

1.9.6 FLOCK HEALTH

The health status of animals is to a large extent a function of the adaptability of the animal, but due to changes in environmental conditions and an increase in host/carriers of disease conditions, health control is primarily a function of management.

An optimal health condition is essential for maximal production and reproduction and is a cost factor on which cuts should not be imposed (see also tables 2 and 3 for vaccination programs for small stock).

1.9.7 RECOMMENDATION TO INCREASE FLOCK FERTILITY

As for all economic characteristics, fertility is determined by:
1. Heredity. The heredity of fertility is very low (±10 %). The reason for this is unfavorable climatic conditions.

2. Environment. The influence of the environment can be ascribed to poor feeding which results in:
   - 2.1.1 Delayed puberty
   - 2.1.2 Shortening of in-heat cycles
   - 2.1.3 Shortening of anoestrus period
   - 2.1.4 Lowering of milk production and mother instinct

It is possible that nutrition shortages (in the form of lick supplements) can occur amongst multiple birth ewes, especially if ewes were on a high level of feeding in the veld, and is then subjected to a low level of feeding in the kraal.

Vit. A should be added during the dry times (Autumn, Winter) of the year to add to the reserves stored in the liver. Supplementation of the following elements should also be considered as they are all, to some extent, important for fertility, prevention of still births, reproduction and multiple births: Vit.E, Selenium, Zinc, Manganese and Iodine.

Make use of breeding seasons which are 6 – 8 weeks long, use small camps with the correct ram / ewe ratio (3/4 rams/100 ewes).

Young ewes that did not conceive for the first time must be culled if there is no other profound reason. In doing so, unproductive animals are removed from the farm before condition is lost at the onset of the dry period.

An effective disease and parasite control program must be part of the flock's management program. The control of external parasites, which can serve as host for harmful diseases, is essential, while control of internal parasites must be done in conjunction with the recommendations of the local veterinarian.

All ewe lambs must be vaccinated with Brucella Rev. I at weaning.

Should multiple births occur, such ewes must receive special care for a period of least three months in order to stimulate milk production.

To overcome problems related to predators, ewes can be placed in small camps or be allowed to lamb on cultivated pastures close to the homestead where they can easily be taken care of.

Losses as a result of predators should be approached genetically through culling mothers with little milk and poor mothering characteristics.

During pregnancy, exposure to a number of factors, such as grazing shortages, poor weather conditions, foot rot, parasite infestations and "suurpens" should be prevented.

These factors could lead to an energy shortage in ewes with two or three lambs, or to those ewes with big lambs. Such shortages lead to

"Domsiek", which usually results in still born lambs. Lambs that are weak at birth, often dies.

Lambs from ewes with energy shortages, tend to grow poorly.

The body mass of ewes should not deteriorate within the 1st 5 weeks of conception as this leads to resumption of the embryo.

Good feeding during the late pregnancy period / mating season, improves milk production after lambing. Maize can be given as creep feeding during the 2 to 3 weeks period before lambing at a rate of 100g/h/d (week – 1, week –2, rest of time 150g/h/d).

1.9.8 CRITERIA FOR FERTILITY IN A EWE FLOCK

Puberty in female animals is determined through the commencement of full sexual activity, that is, the production of ovum and signs of being in heat. The average age, at which this takes place, is between 9 and 11 months in small stock. Sexual maturity in a ewe represents the age at which she can be mated successfully without negative effects on her growth and future reproduction. This will therefore be after the puberty stage in ewes. Sexual maturity varies between breeds.

Number of lambs produced during the ewe's lifetime. Age at first lambing – The earlier a ewe lambs, the better the chances that she will produce more lambs in her lifetime. Too early lambing however, can stunt growth and development of the animal as a result of the negative correlation between the release of sex hormones in the animal's body and further bone growth and development. A target mass for the first mating seems to be a more acceptable criteria, for example first mating takes place on 10 – 12 months of age (two teeth) with a body mass of 28 – 35 kg (varies between breeds).

Number of mating or inseminations per conception – Should a mating period include a maximum of two estrus cycles, then ewes which did not ovulate, will not get pregnant and will automatically be culled.

Selection for fertility amongst rams are more important, as a ram with low fertility would not only decrease lambing percentage, but would also lower the lambing percentage of the flock.

1.9.9 FEBRUARY

Stop mating in the middle of the month.
Give the ewes a well balanced formulated lick during their pregnant period.

1.9.10 MARCH

Give the ewes a well balanced formulated lick during their pregnant period.

1.9.11 APRIL

Give the ewes a well balanced formulated lick during their pregnant period. (150 days)

1.9.12 MAY

Give the ewes a well balanced formulated lick during their pregnant period.
Deworm ewes, and treat against external parasites.
Prepare for the lambing period. (6 weeks)

1.9.13 JUNE

Lambing season starts.
Maize can be given as creep feeding during the 2 weeks before / during lambing season, 100g/h/d. – week – 1, week –2, rest of time 150g/h/d up to weaning of lambs at 90 days.

1.9.14 JULY

Lambing period. (6 weeks)
Ewes and lambs must be in camps with good grazing and a balanced formulated lick.

1.9.15 AUGUST

Ewes and lambs must be in camps with good grazing and a balanced formulated lick.
Deworm lambs – milk tape worms, treat against external parasites.

1.9.16 SEPTEBER

Wean lambs – 90 / 120 days.
Ewes rest phase for ± 4 weeks.
Ram – Ewe lambs must be vaccinated with Brucella Rev 1, Multivax P (2x),
Bluetongue (1-2-3), Live Rift Valley Fever and Wesselsbron Disease after weaning.
Move ewes and lambs to camps with good grazing and a balanced formulated lick.The distance between camps must be ± 5km.
Put 5/6 old ewes or rams (castrate) with the lambs in same camp to guide them.

1.9.17 OCTOBER

Prepare for mating season 30/35 days after weaning.
Mating starts for 42 days.
Vaccinate ewes, rams one month prior to mating – Multimin – E to prevent shortages of minerals and vitamins, and to increase fertility.
Deworm ewes, rams, and treat against external parasites.
On the day of mating start vaccinating rams with Multimin – E to prevent shortages of minerals and vitamins, and to increase fertility.
Good grazing and a balanced formulated lick must be available.
Maize can be given as creep feeding during the 2 weeks before mating season begin.

1.9.18 NOVEMBER

End of November, mating season stops after 42 days.
Give the ewes a well balanced formulated lick during their pregnant period.

1.9.19 DECEMBER

Give the ewes a well balanced formulated lick during their pregnant period.

1.9.20 JANUARY

Give the ewes a well balanced formulated lick during their pregnant period. (150 days).
Prepare for the lambing period. (6 weeks)

1.9.21 FEBRUARY

Give the ewes a well balanced formulated lick during their pregnant period.
Prepare for the lambing period. (6 weeks)
Deworm ewes, and treat against external parasites.
Maize can be given as creep feeding during the 2 weeks before lambing season begin / during 100g/h/d. – week – 1, week –2, rest of time 150g/h/d up to weaning of lambs at 90 days (Depends on grazing after the rains)

1.9.22 MARCH

Lambing period (6 weeks)
Maize can be given as creep feeding during the 2 weeks before lambing season begin / during - 100g/h/d. – week – 1, week –2, rest of time 150g/h/d up to weaning of lambs at 90 days (depends on grazing after the rains).

Give the ewes a well balanced formulated lick during their lambing period.

1.9.23 APRIL

Lambing period.
Maize can be given as creep feeding.
Ewes and lambs must be in camps with good grazing and a balanced formulated lick.

1.9.24 MAY

Ewes and lambs must be in camps with good grazing and a balanced formulated lick.
Deworm lambs – milk tapeworms treat against external parasites.

1.9.25 JUNE

Wean lambs – 90 / 120 days.
Ewes rest phase for ± 4 weeks.
Ram – Ewe lambs must be vaccinated with Brucella Rev 1, Multivax P (2x), Bluetongue (1-2-3), Live Rift Valley Fever and Wesselsbron Disease after weaning.
Move ewes and lambs to camps with good grazing and a balanced formulated lick.
The distance between camps must be ± 5km.
Put 5/6 old ewes or rams (castrate) with the lambs in same camp to guide them.

1.9.26 JULY

Prepare for mating season 30/35 days after weaning.
Mating begin for 42 days.
Vaccinate ewes, rams one month prior to mating – Multimin – E to prevent shortages of minerals and vitamins, and to increase fertility.
Deworm ewes, rams, and treat against external parasites.
On the day of mating start, vaccinate rams with Multimin – E to prevent shortages of minerals and vitamins, and to increase fertility.
Good grazing and balanced formulated lick must be available.
Maize can be given as creep feeding.

1.9.28 AUGUST

Mating stops after 42 days.
Give the ewes a well balanced formulated lick during their pregnant period.

1.9.29 SEPTEMBER

Give the ewes a well balanced formulated lick during their pregnant period.

1.9.30 OCTOBER

Give the ewes a well balanced formulated lick during their pregnant period.

1.9.31 NOVEMBER

Give the ewes a well balanced formulated lick during their pregnant period.
(150 days)
Prepare for the lambing period. (6 weeks)
Maize can be given as creep feeding during the 2 weeks before lambing season begin / during - 100g/h/d. – week – 1, week –2, rest of time 150g/h/d up to weaning of lambs at 90 days (depends on grazing after the rains).
Deworm ewes, and treat against external parasites.

1.9.32 DECEMBER

Lambing season starts.
Maize can be given as creep feeding during the 2 weeks before lambing season begin / during - 100g/h/d. – week – 1, week –2, rest of time 150g/h/d up to weaning of lambs at 90 days (depends on grazing after the rains)
Ewes and lambs must be in camps with good grazing and a balanced formulated lick.

CHAPTER 2: MANAGEMENT FOR LOW LAMB MORTALITY

2.1 CARE OF LAMBS

The ideal would be for the ewes to lamb close to home. Ewes that experience difficulty in giving birth, must be assisted. Sometimes the lambs must be repositioned. The forelegs and head usually emerge first, with the lamb's back turned towards the ewes back. Ewes should only be helped if it is absolutely necessary; otherwise they should be left to lamb completely on their own. Most mortalities in small stock farming occur during the lambing season; therefore special attention must be given to ewes and lambs.

Generally 10 to 15 % of mortalities occur at birth or the first few days thereafter. Poor conditions, multiple birth and poor management can be some of the reasons for this. Lambs can be put in a pen shortly after been born until they know their mothers before they can accompany them to the camps.

However this practice has many disadvantages. It requires a lot of time and labour. Every evening when the ewes return from the veld, the farmer must see to it that every ewe is united with her lamb. The farmer must also make sure that the lambs are fed during the night, before the ewes go out for grazing in the morning.

The lambs in the pens must get fodder to help them feed on their own. The pens can often be infected with disease carrying organisms, lambs also get infected and disease can spread rapidly. It is better to leave the lambs with their mothers in small camps for approximately two weeks. Other group ewes and lambs can then be placed together with this group. Ewes with sturdy lambs should be moved to larger camps on a weekly basis.

Lambing camps should be inspected daily to make sure those weaker lambs are being fed. It is essential that all lambs should have colostrums within two hours after birth, if the mother cannot provide it, lambs should drink at other ewes. Artificial colostrums can also been given to make them more resistant to certain diseases. Selection must be strictly applied against ewes with "calabash" likes teats which makes it difficult for the lambs to drink, as well as sisters teats, or ewes abandoning their lambs.

Where triplets occur, it must be established whether the ewe has sufficient milk to raise them. If not, one lamb should be given to a ewe with only one lamb, or a ewe whose kid/lamb has died. If another ewe doesn't accept the lamb, the lamb must be hand-reared.

The same applies in the case of quadruplets. If ewes have enough milk to raise a triplet, the ewe with the lambs must be placed in a separate camp where supplementary feed can be provided to them.

2.2 LAMB MORTALITIES

2.2.1 GENERAL

Investigation in most of the small stock production countries indicated that between 15 and 20 percent of the lambs born, die between birth and weaning age. Most of the lambs are stillborn or die within three or four days after birth. Inherited infections may be responsible for outbreaks of abortion.

According to an official survey, an average of 15 percent of the lambing crop in Namibia die in the period between birth and weaning age. A conservative estimate puts this figure at about 126 000 lambs.

These losses in terms of money are alarming:

- Supplement feed to pregnant ewes was wasted.

- Ram cost per lamb increase for the remaining lambs.
- A decline in breeding progress due to the loss of future breeding material.
- Where applicable, the loss of secondary products, such as wool mass in pregnant ewes.

TABLE 2.1 MAIN CAUSES OF LAMB MORTALITIES UP TO WEAN (%)

STILL BORN	DISEASES	LITTLE MILK	PREDATORS	OTHER
43.4	15.4	10.4	14.8	16.0

In a study conducted in West Australia it was found that malnutrition after birth, difficult birth and small or weak lambs were responsible for 90 % of all lamb mortalities. Veterinary Services in Namibia (Annual Report 1992), support these findings.

2.2.2 BODY RESERVES

Under normal conditions, the newborn lamb has sufficient energy reserves in the form of body fat to survive for two or three days. When these reserves are however depleted, the lamb died. In order so survive, the lamb must have colostrums or milk as soon as possible after birth.

2.2.3 BIRTH MASS

The ideal birth mass for lambs is about 3.5 to 4.1 kg. Lighter lambs are usually weak and have a slimmer chance of survival, while heavier lambs cause difficult births that can also result in mortalities.

TABLE 2.2 LAMB MORTALITIES FROM BIRTH TO WEAN AGE IN THE DIFFERENT WEIGHT GROUPS

MASS GROUP	MORTALITY PERSENTAGE
0.90 – 1.35	70
1.35 – 1.80	65
1.80 – 2.25	40
2.25 – 2.70	30
2.70 – 3.15	17
3.15 – 3.6 0	9
3.60 – 4.05	4
4.05 – 4.50	10
4.50 – 4.95	20
4.95 – 5.40	50

From the group with a birth mass of less than 1.80 kg, 65 % died, between 1.80 and 2.7 kg about 35 % died and between 3.60 and 4. 5 kg only 4 %. As die birth mass increases, the mortality percentage above 4.05 kg rises again.

2.2.4 LITTER SIZE

The interaction between litter size and the age of the ewe has proved that twins of five-year old ewes have a better chance of survival than singles of two-year old ewes. These results indicate intensive management during lambing – time and for the first three days after birth. Where triplets occur, it must be established whether the ewe has sufficient milk to raise them. If not, one lamb should be given to a ewe with only one lamb, or a ewe whose lamb died.

2.3 STATEGIES FOR IMPROVED LAMB SURVIVAL

Starvation – mismothering – exposure complex causes approximately 50 % of all deaths.

- Improve nutrition and management.
- Before mating, cull ewes with poor constitution and with teeth problems.
- Cull all ewes with defective udders (mastitis, calabash teats).
- Cull all ewes that abandon their lambs.
- Vaccinate pregnant ewes enterotoxaemia, lamb dysentry and lock jaw.
- Pregnanted ewes must not be vaccinated agianst Blue tongue.
- From a month to at least two weeks before lambing, get the flock used to torches or gas lamps at night, the idea is that the ewes must become used to your presence at night
- Use pregnancy scanning to divide the flock in single- and twin-bearing ewes. Scan ewes 42 to 60 days after the rams have been remove from the ewes.
- Complete all dosing, dipping, shearing, hoof trimming and vaccination 4 to 6 weeks prior to the start of the lambing season.
- Ensure that the ewes get the best feed available during the last 6 weeks of pregnancy and the first 6 to 8 weeks of lactation. This ensures strong, viable lambs and good milk production.
- As soon as a ewe lambs, check to see if she has milk and whether or not the lamb can suckle. If the ewe does not have milk, the lamb must be taken away and the ewe marked for later sale.
- Do not lamb ewes in paddocks at a density of more than 18 ewes/ha, younger ewes are not quite as good mothers as the mature ewes and require more attention.
- Ensure the lambing paddock is protected against cold weather, shelters in the form of bushes or long grass protect against sun and cold winds.
- Foster lambs with ewes with single lamb, or those who have lost their lamb. The process is made easier by smearing the foster lamb with the afterbirth or fluids of the ewe.

- Obtain colostrums from ewes which have lambed recently and store it frozen. This can then be warmed to body temperature (39 °C) and dosed to newborn lambs at 50 ml / kg.
- Vaccinate lambs against the relevant diseases that occur. (Enterotoxaemia)
- Dose lambs regularly against tapeworm before weaning.
- Give special attention to multiple births.
- Change the lambing season if losses related to unfavorable weather conditions are high during the traditional lambing season.
- Do not allow overcrowding of ewes in lambing paddock.
- Too much frequent movement of ewes with newborn lambs, doesn't allow the bond between ewe and lambs to establish properly.
- Overcrowding during nightly kraaling could cause a significant increase in lamb deaths.
- Make use of two lambing seasons if effective supervision is not possible in one lambing season.
- Artificial colostrums can be made up by:
  = 500 ml cow milk,
  = 1 egg beaten into milk,
  = 1 tea spoon sugar,
  = 1 tea spoon cream/cooking oil/fish oil

2.4 REPLACEMENT EWES LAMBS

Over –nutrition may cause oversized lambs, which the ewe cannot expel easily at birth. This may cause the death of the ewe, the birth of a stillborn lamb or a live lamb which, due to brain damage during the protracted birth process, cannot suckle properly and dies with in a few days.

2.5 WEANING OF LAMBS

The age at which lambs are weaned, will depend on their age and weight. It must be kept in mind that milk production of a ewe starts to decrease after about six weeks. Lambs should be weaned at the age of 3 to 4 months. Where creep feeding is provided it may be possible to wean lambs at an earlier age.

Lambs that are, however, weaned too early, suffer from stunted growth and become much older before attaining the desired age for mating or marketing. It is better to withdraw the ewes from the lambs and leave the lambs in the original camp. The lambs are then much calmer and forget their mothers quicker, especially if a few mature ewes are placed with them to guide them. Ram and ewe lambs must not be kept together in the same camp for more than four months, as the ram lambs become fertile.Weaners must be provided with good grazing in order to grow fast. Lambs that were weaned are very sensitive to endo-and ectoparasites and close attention must be given to any signs of infestations.

CHAPTER 3: THE BASIC PRINCIPLES OF SELECTION OF BREEDING STOCK

3.1 INTRODUCTION

The only way a livestock breeder can change the genetic composition of his animals is by ensuring that preferred traits are transmitted from one generation to the next. He can do this either by selecting specific animals for his breeding programs or by practising a specific mating method e.g. line breeding and crossbreeding.

3.2 SELECTION

The aim of selection is two fold:

- First, to increase production in the current flock by selecting superior animals, which will maintain a large proportion of their production superiority during the rest of their lifetimes.
- The second and more important aim is to increase production in the following generations by selecting parents whose offspring will exhibit a large proportion of their combined superiority.

Selections can be divided into natural and artificial selections. In the natural selection, the main force responsible for selection is the survival of the fittest in a particular environment. Artificial selection is the process by which man decides which animals in a generation will be allowed to become parents of the next generation and how many progeny they will be permitted to have.

3.3 SELECTION OF BREEDING RAM / EWE

3.3.1 INTRODUCTION

It is advisable that one personally purchases the ram. One should insist on obtaining the records from the breeder. If the breeder cannot produce records, another breeder should be tried. Background knowledge on the breeder should be obtained, for example the size of his flock.

3.3.2 RECORDS NEEDED TO PURCHASE A RAM/ EWE

- Birth-mass, - birth problems.
- Forty-two day mass – indicates the mother's milk production.
- Weaning mass and index.
- Six – month index.
- Twelve – month growth index.
- Records of father and mother

3.3.3 CULL FAULTS

A functional and efficient ram/ewe is required, so start selection by culling for:

- Feet (outgrowing or abnormal hoofs are cull faults)
- Legs (crooked, bandy legs and sickle hocks)
- Colour (face, ears, legs, nose and horn)
- Mouth (over and under-shot jaws)
- Scrotal circumference – 30 cm
- Conformation faults.
- Phenotype – breed standards.

THE SELECTED RAM MUST LOOK LIKE A RAM AND NOT A EWE, OR A EWE LIKE A EWE AND NOT LIKE A RAM.

CHAPTER 4: ELEMENTARY CONCEPTS OF SMALL STOCK NUTRITION – SHEEP AND GOATS

4.1 INTRODUCTION

The aim of a small stock farmer is to obtain the maximum output per hectare at an economical input cost. The feed cost of a sheep/goat enterprise is approximately 65 to 72 % of the total allocated cost. It is important for the small stock producer to know the basic requirement of sheep so that he can use these correctly and as profitably as possible during the different production stages of the animal. There are two categories of nutritional requirements, - requirement for maintenance, and requirement for production. The nutrient requirement of small stock is determined largely by factors such as body size (mass), physiological state and environmental conditions e.g. intensive or extensive, warm or cold weather. The body mass or size of an animal has a significant influence on the nutritional requirements of the animal. If the body mass is in the first instance seen as a measure of maturity that young, growing animals have reached, body mass has a dramatic influence on the nutritional requirements.

4.1.1 ENVIRONMENTAL INFLUENCES

It is essential to consider other aspects that also exert considerable influence on the nutrient requirements of the grazing sheep. The first aspect is the energy lost as a result of muscular action. Researchers have found that the energy requirements of a grazing animal are 10 to 20 % higher than those of a penned animal. The amount of E used depends largely on the availability of grazing. Poor veld conditions and large camps force animals to walk long distances, and this increases the amount of E used. A second important factor under grazing conditions is the weather, temperature, rain and wind. A combination of rain and wind, coupled with low temperatures can cause a drastic increase in nutrient requirement of small stock. High temperatures appear to be more important than low temperatures and may cause an even greater stress, which may interfere with grazing patterns and causes a reduced intake. The nutrients required by animals are water, protein, lipids, minerals, and vitamins.

4.1.2 WATER

Water is an extremely important compound (nutrient) for livestock because it makes up 71-73 % of the fat free animal's body weight. It has numerous functions of vital importance and in general, is the most vital material ingested by animals, since a lack of water will have more immediate and drastic effects on animal physiology than the lack of any other nutrient.

Water has many functions. It acts as a solvent for many different compounds. It serves to transport fluids and semi-solids ingested through the GI-tract. It transports materials in the blood and other body tissues so that nutrient is moved to the cell and waste away from the cells. It serves in the elimination of the body wastes via urine and, lastly is used for evaporative cooling of the animal's body when temperatures are elevated.

Water is lost from the body by ways of the kidneys as urine, from the GI- tract in the faeces, from the lungs and skin as water vapor, and from the sweat glands as sweat. Water quality, as well as quantity, may affect feed consumption and animal health.

4.1.3 PROTEIN

Proteins are essential constituents of the tissues of all biological organisms. In animals, proteins are found in higher concentration in organ and muscular tissue than any other constituent, except water. All cells synthesize proteins for part or all of their life cycle and without protein synthesis life could not exist. Proteins are large molecules with molecular weights ranging from 35 000 to several hundred thousand.

Each protein has a distinctive function in the animal's body (or other biological organisms), ranging from protection of the body surface (hair, skin) to defense against invading organisms. Structurally, proteins have important functions as components of muscle cell membranes, skin, hair, and hooves.

Metabolically important proteins are the blood serum proteins, enzymes, hormones and immune antibodies, which all have important specialized functions in the body. Proteins are synthesized in plants and animal cells where the cell nucleus contains genetic material that determines the nature of the newly synthesized protein. The genetic material commonly referred to as DNA (deoxyribonucleic acid), is transferred from one generation to another. The nuclear DNA controls the synthesis of all proteins regardless of their function. Thus proteins are vital to animals and provided in the diet in one from or another. Requirements are always highest (in terms concentration in the diet) for young, rapidly growing animals. The needs decrease as the growth rate declines. Requirements are lowest for adult animals in a maintenance situation. They are increased during pregnancy and increased markedly during periods of peak lactation or egg production. Protein deficiency can be a result of one or more limiting amino or inadequate protein consumption. Signs of protein deficiency include poor growth rate and reduced N retention by the body, poor utilization and lower consumption of feed, lowered birth weights often accompanied by high infant mortality, reduced milk or egg production and infertility in male and females.

4.1.4 CARBOHYDRATES

Although plants synthesize many different carbohydrates, the basic compound is glucose, from which more complex or different carbohydrates are synthesized. In plant tissues, carbohydrates may comprise 50 % of the dry matter of forages and as much as 80 % in the kernels of some cereal grains. Thus carbohydrates are the major dietary components for all herbivorous animals. For the animal, carbohydrates serve as a source of E or as bulk in the diet, but there is no specific requirement for any individual carbohydrate compound. Glucose (also called dextrose) and fructose are the most common simple sugars in feed and food ingredients. They occur as the simple sugars in both plants and tissues, but only in low concentrations. Fructose is converted readily to glucose in the animal body and is therefore available to body metabolism as glucose. A high proportion of simple sugars will be converted to glucose in the wall of the small intestine. Those that are not, will be modified by the liver or metabolized in other ways. Thus, all absorbed sugars become available to the body cell for E or other metabolic.

4.1.5 LIPIDS

Lipids are organic compounds that are insoluble in water but soluble in organic solvents. Quite a variety of different types of compounds are found in both plants and animal tissues, all of which serve some important biochemical or physiological function. Chemically, lipids range from fats and oils to complex sterols Lipoproteins and are important constituents of all cell structures. Fats serve as a concentrated form of stored energy.

4.1.6 MINERALS

With respect to animal nutrition, the minerals that are dietary essentials are classified as the major or macro-minerals and the trace or micro-minerals. The major minerals are normally present in animal's carcasses at levels greater than 100 ppm. Included in this group are calcium (Ca), chlorine (CL), magnesium (Mg), phosphorus (P), potassium (K), sodium (Na), and sulfur (S). Ten trace minerals are usually present in the carcass at levels less then 100 ppm. Included in this group are chromium (Cr), cobalt (Co), copper (Cu), fluorine (F), iron (Fe), iodine (I), manganese (Mn), molybdenum (Mo), nickel (Ni), selenium (Se) silicon (Si), and zinc (Zn). The most obvious function of mineral elements in the body is to provide structural support in the form of the skeleton. Bone is formed through the deposition of Ca and P in a complex salt in a protein matrix. Mineral elements are absorbed from the GI tract by either an active or passive method. Mineral elements are absorbed primarily in the ionic form. A deficiency of Ca or P or Vit. D or an imbalance may result in rickets in young animals, which is manifested by inadequate mineralization of the bone, crooked legs, and enlarged joints as well as other abnormalities. In older animals the minerals are withdrawn from the bones, resulting in osteoporosis. With regards to Na, it is widely recognized that livestock need added salt, when salt is not fed and the soil or water supplies do not contain much Na, deficiencies may result. Clinical signs are a craving for salt, emaciation, listlessness, and poor performance. A deficiency of K is not likely except in animals fed very high grain diets. Deficiency signs are similar to those of Na.

With regards to the trace minerals, Fe is always deficient for young pigs because body reserves of newborn pigs are low. A Co deficiency results in a deficiency of Vit. B_{12}, because Co is an essential constituent of this vitamin. Animals appear listless and will develop a particular type of anemia. Typical signs are a light hair color, partial paralysis of the rear quarters. Se is deficient in many areas for domestic livestock. A clinical deficiency shows up as white muscle disease, primarily in young animals. Zn deficiencies are also relatively common and can be made more severe by high levels of Ca consumption. One sign is parakeratosis which is dermatitis manifested by itching, skin lesions.

4.1.7 VITAMINS

Vitamins are organic substances that are required by the animal's tissues in very small amounts. All vitamins are essential for animal tissues, but some species of animals are able to synthesize certain vitamins in their tissues or they are able to utilize vitamins synthesized by micro organisms in their GI tract. The primary function of many water-soluble vitamins is as coenzymes. Vitamin A, is concerned with vision and with maintenance of epithelial cell, (cell which line body cavities and cover body surfaces). Vitamin D important in absorption and metabolism of Ca and in bone metabolism. Vitamin E function as a metabolic antioxidant and vitamin K is concerned with the blood clotting mechanism. The major storage site for most vitamins is the liver, with less in the kidney, spleen, and other tissues or organs. Most are stored bound to specific proteins. Vitamins stored in the tissues are released at a rate necessary to maintain a relatively constant level in the blood. Vitamins are absorbed primarily from the small intestine. The B-complex vitamins like vitamin K is synthesized by micro organisms in the testiness of monogastric species and in the rumen of ruminating animals. Many of the signs of the various vitamin deficiencies are similar. Those include anorexia (poor appetite), reduced growth, dermatitis, weakness, and muscular incoordintion. For example, vitamin A deficiency can cause various kinds of blindness, including night blindness and total blindness.

4.1.8 ENERGY

Quantitatively, energy is the most important item in an animal's diet, and all feeding standards are ration formulation based on some measure of energy with additional inputs on protein or amino acids, essential fatty acids vitamins and minerals. Energy is often a limiting dietary component for free-ranging animals, but it would not, normally be a problem for confined animals raised under intensive situations. As a ruminant the sheep is well equipped in symbiosis with the microbe population in the rumen, to utilize plant material as primary source of energy. The object with supplementary energy, usually some or other form of maize, is thus to try to compensate for this decrease in voluntary nutrient intake which sheep normally experience on natural winter pastures. If the quantity of supplementary energy provided is too small, it is in any case naive to expect any significant reaction. A factor which has a significant influence is the energy loss due to muscle movement . Research has shown that the energy requirement of animals on natural grazing is approximately 10 to 20 times higher than those of animals kept in kraals. The loss of energy depends to a large extent on the availability of grazing and thus the ease with which animals can graze. Poor grazing and too large camps, where the animals have to walk great distances per day, contribute to an increased energy loss. A second factor which plays a significant role under grazing conditions, is the prevailing climatic conditions such as temperature, rain and wind.

4.2 BODY MASS

The body mass or size of a sheep/goat has a marked influence on the nutrient requirement of animals. (Table 4.1)

TABLE 4.1 THE INFLUENCE OF BODY SIZE ON THE NUTRIENT REQUIREMENT OF YOUNG REPLACEMENT EWES *(NRC, 1975)*

BODY MASS (W)	ADG	DM	INTAKE /DAY	CP	TDN	ME	Ca	P
KG	G/DAY	KG	(% OF W)	(G/DAY)	(KG/DAY)	(MJ/DAY)	(G/DAY)	(G/DAY)
30	180	1.3	4.3	130	0.81	9.3	5.9	3.3
40	120	1.4	3.5	133	0.82	9	6.1	3.4
50	80	1.5	3	133	0.83	8.25	6.3	3.5

ADG = Average daily gain **DM** = Dry matter
P = Phosphorus **Ca** = Calcium
CP = Crude protein **ME** = Metabolic E expressed in mega joules/kg
TDN = Total digestible nutrients

4.3 NUTRITIONAL REQUIREMENT OF THE RAMS, EWES AND LAMBS

The young ram that is still actively busy to grow and develop has a considerable higher requirement of nutrients when compared to more mature rams. It is also important to take notice of the fact that the nutritional requirements of rams are at all times higher than those of dry ewes of comparable body mass.

The nutrient requirement of ewes increase during the different reproduction stages. During early pregnancy there is only a small increase in the nutritional requirement of a ewe above that of a dry ewe. During the last 6 weeks of pregnancy the ewe experiences a high increase in nutritional requirements by growing the fetus. In the case of a young pregnant ewe, busy to grow herself, this requirement for nutrients will be even higher. The better the nutrition, the higher the percentage of multiple births. During lactation ewes experience a further increase in nutritional requirement, this increase is however closely correlated to the level of milk production of the ewe. This is especially important in the case of multiply births.

In relation to their body mass, young lambs have the highest nutritional requirements. The younger the lamb the higher its requirement, especially for high quality proteins. It is also important to take notice of the fact that a young lamb is busy to develop into a full-fledged ruminant. The lamb starts its life as a typical single stomach animal (non-ruminant) which is adjusted to the digestion and utilization of a milk diet. Roughage stimulates the development of the reticulum and consequently the lamb develops into a full- fledged ruminant. This transition from a non-ruminant to a ruminant, starts at 3 weeks after birth. For this reason the fodder that lambs receive must be of a high quality as the lamb are at the stage of their development highly susceptible.

TABLE 4.2 THE NUTRIENT REQUIREMENT OF EWES *(NRC 1975)*

PHYSIOLOGICAL	BODY MASS	DM	INTAKE DAY	CP	TDN	ME	Ca	P
STATE	Kg	Kg	(% of W)	(kg/DAY)	(MJ/DAY)	(G/DAY)	(G/DAY)	(G/DAY)
Maintenance	50	1.0	20.	89	0.550	8.25	3.0	2.8
	60	1.1	1.8	98	0.610	8.25	3.1	2.9
	70	1.2	1.7	107	0.660	8.25	3.2	3.0
	80	1.3	1.6	116	0.720	8.25	3.3	3.1
Pregnant	50	1.1	2.2	99	0.600	8.25	3.0	2.8
First	60	1.3	2.1	117	0.720	8.25	3.1	2.89
15 Weeks	70	1.4	2.0	126	0.770	8.25	3.2	3.0
	80	1.5	1.9	135	0.820	8.25	3.3	3.1
Pregnant	50	1.7	3.3	158	0.990	8.70	4.1	3.9
Last	60	1.9	3.2	177	1.100	8.70	4.4	4.1
6 Weeks	70	2.1	3.0	195	1.220	8.70	4.5	4.3
	80	2.2	2.8	205	1.280	8.70	4.8	4.5
Lactating	50	2.1	4.2	218	1.360	9.75	10.9	7.8
First	60	2.3	3.0	239	1.500	9.75	11.5	8.2
8 Weeks	70	2.5	3.6	260	1.630	9.75	12.0	8.6
Single lamb	80	2.6	3.2	270	1.690	9.75	12.6	9.0
Lactating	50	2.4	4.8	279	1.560	9.75	12.5	8.9
First	60	2.6	4.3	299	1.690	9.75	13.0	9.4
8 Weeks	70	2.8	4.0	322	1.820	9.75	13.4	9.5
Twin lambs	80	3.0	3.7	345	1.950	9.75	14.4	10.2

ADG = Average daily gain **DM** = Dry matter
P = Phosphorus **Ca** = Calcium
CP = Crude protein **ME** = Metabolic E expressed in mega joules/kg
TDN = Total digestible nutrients

4.4. FLUSHING

If flushing is successful, it will result in the production of more ova, improved conception and implantation of the ova in the uterus, a higher conception rate and ultimately a higher lambing percentage. Flushing of ewes should start 21 days before mating, and possibly continue for 21 days into the breeding season. Flushing can take several forms, according to circumstances:

- moving to fresh pasture
- reducing stocking rates
- feeding a legume hay
- feeding maize silage
- creep feeding (high E rate of 200/450 g/day) when good pasture is scarce.

4.4.1 FEEDING OF LAMBS

TABLE 4.3 THE NUTRIENT REQUIREMENT OF LAMBS *(NRC, 1975)*

BODY MASS (W)	ADG	DM	INTAKE /DAY	CP	TDN	Ca	P
KG	G/DAY	KG	(% OF W)	(G/DAY)	(KG/DAY)	(G/DAY)	(G/DAY)
Early weaning							
10	550	0.6	6	96	0.44	2.4	1.6
20	275	1	5	160	0.73	3.6	2.4
30	300	1.4	4.7	196	1.2	5	3.3
Fattening							
30	200	1.3	4.3	143	0.83	4.8	3
35	220	1.4	4	154	0.94	4.8	3
40	250	1.6	4	176	1.12	5	3.1
45	250	1.7	3.8	187	1.19	5	3.1

ADG = Average daily gain
Ca = Calcium
TDN = Total digestible nutrients
DM = Dry matter
CP = Crude protein
P = Phosphorus

4.4.2 FEEDING BEFORE WEANING

Young lambs in relation to their body mass, have the highest nutrient requirement. Lambs obtain their nutrients from three sources:

- Their mother's milk
- Creep feeding
- Grazing

Milk is essential in the first 3 to 4 weeks of the lamb's life, and during this period the correlation between milk intake and live mass gain is very high, i.e. more milk lamb consumes more mass it will gain.

4.5 CREEP FEEDING

Definition: is a balance ration that is supplied to un-weaned lambs and to which the ewes have no access.

- Creep feeding is not a practice that can solve all feed and management problems.
- The major objective of creep feeding is to supplement E.
- Non- protein nitrogen (NPN) should be avoided as far as possible by young animals.
- It is a sound practice to include a minimum of 15 % roughage in a ration.
- Lambs should be introduced to creep ration at 3 weeks of age.
- Creep feeding is always given to lambs *ad lib.*
- Creep feeding must be palatable to achieve intake.

Creep feeding can achieve the following advantages:

- Lambs can be weaned early.
- Ewes can more quickly regain mass lost.
- Lamb learns to eat earlier.
- Rumen function is stimulated earlier.
- Weaning shock is reduced.

4.5.1 FEEDING THE RAM

TABLE 4.4 THE NUTRIENT REQUIREMENT OF RAMS *(NRC, 1975)*

BODY MASS (W)	ADG	DM	INTAKE /DAY	CP	TDN	Ca	P
KG	G/DAY	KG	(% OF W)	(G/DAY)	(KG/DAY)	(G/DAY)	(G/DAY)
40	**250**	**1.8**	**4**	**184**	**1.17**	**6.3**	**3.5**
60	200	2.3	3.8	219	1.38	7.2	4
80	150	2.8	3.5	249	1.54	7.9	4.4
100	100	2.8	2.8	249	1.54	8.3	4.6
120	**50**	**2.6**	**2.2**	**231**	**1.43**	**8.5**	**4.7**

ADG = Average daily gain **DM** = Dry matter
Ca = Calcium **CP** = Crude protein
TDN = Total digestible nutrients **P** = Phosphorus

4.5.2 YOUNG RAMS

In the nutrition of young rams, provision should be made for the normal development of the body, which is mainly development of muscle (protein) and bone (Ca and P) and not for the deposition of fat.

4.5.3 BREEDING SEASON

For a ram to effectively work during the mating season, rams should not be too fat. Supplementary feeding should begin six weeks prior to mating. With adult rams a concentrate supplement of 0.2 to 0.5 kg can be given with good pastures, example – 40 % maize, 40 % oats (grain or roll) and 20 % wheat bran.

CHAPTER 5: GENERAL MANAGEMENT OF SMALL STOCK

5.1 INTRODUCTION

Management on any farm is an all-encompassing term, as any decision-making is a part of management. It includes various aspects such as feeding, breeding, disease control, capital investment, keeping of records, manpower utilization as well as the handling and care of flock.

5.2 DOCKING OF TAILS

The ewes of some of the fat-tailed breeds tent to develop large heavy tails. Although the rams acquire a certain technique by which the tail is pushed away with their shoulder, it often happens that the ram wastes too much energy in his efforts to service a ewe with a large tail. In some mutton breeds like Dorper, the docking of the tail of ewes as well rams are a common phenomenon due to Breeding standards. The easiest and most common method is the use of the rubber ring. The cutting-off method used earlier is a complicated operation and is not commonly used any longer.

5.3 CASTRATION

The methods that can be used for the castration of male lambs are the knife method, rubber ring and burdizzo. The burdizzo method is more often used in the case of mature rams. Normally farmers castrate their lambs at weaning age. Before and after castration, lambs must not be chased around.

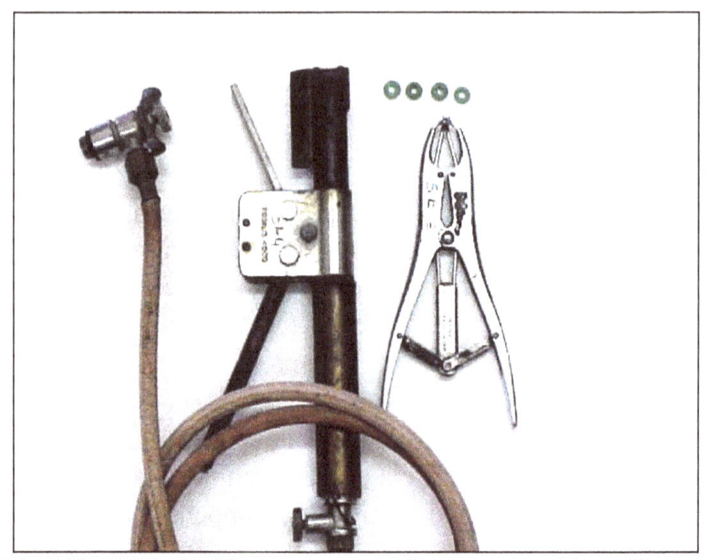

Fig. 5.1 Gas Tail Cutter and Castrator Applicator and rubber rings.

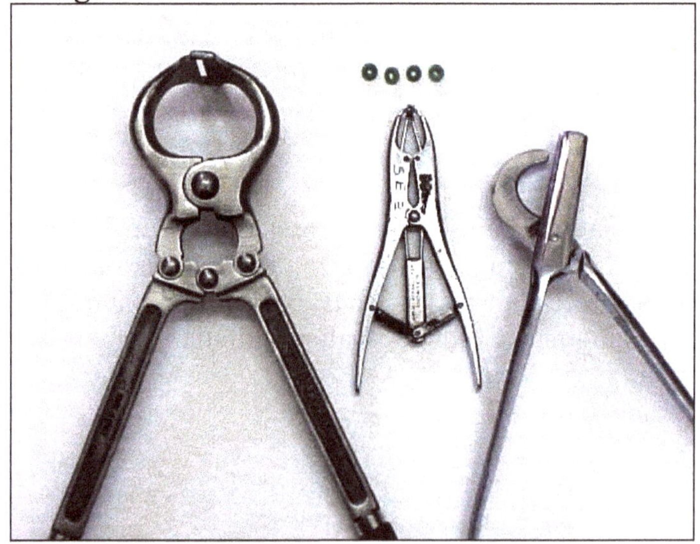

Fig. 5.2 Burdizzo castrator, Castrator Applicator and rubber rings and Testicle Applicator.

5.4 SHEARING

5.4.1 INTRODUCTION

In woolly breeds the same constraints with multi-sire mating groups apply as with mutton sheep. Wool is essentially produced for export and it is therefore subjected to world trade and market trends. The price differences between fine and medium fibres influence trends in breeding and selective correlation between fibre diameter and fleece weight. In the wool stud industry it seems desirable to fix and maintain certain traits in specific lines or flock in environments where they are well adapted.

Sheep like Karakul sheep and Merino sheep are normally shorn twice per year, via February/April and again in September/October. The shearing of other breeds like the Dorper sheep depends on the farmer. Lambs are shorn the first time at the age of 6/7 months, depending on the length of their wool and the danger of getting stuck in bushes. Lambs are very sensitive to rain after being shorn and should have shelter at night. Rams must receive special attention and must be not be unnecessarily cut. It is desirable to shear the rams first, then the young animals and lastly the old animals. The shearing shed and all the equipment used during shearing should be disinfected every day with a disinfectant that contains phenol (5%). Wounds caused by shearing must immediately disinfected and treated with wound oil. During the time the animals are shorn, the animals can also be treated for ticks and internal parasites and lost ear tags can also be replaced.

5.4.2 SHEARING SHEDS

In practice it has been found that designing a shearing shed and its associated facilities can be a complex team-project. Regardless of the size of the sheep enterprise, the shearing facilities need to be planned to cater for the following:

- A simple and effective sheep flow.

- Good light for shearing and wool handling.
- A simple and logical layout of wool handling equipment from shearing area to baler.
- Adequate bale storage place.
- Orientation of the building, takes into account ventilation, light and exclusion of rain.
- Situated near sheep-handling yard, if possible.

Fig. 5.3 Shearing of sheep, wool bales and sorting wooden boxes

5.4.3 MARKETING WOOL, MOHAIR AND KARAKUL PELTS

Wool and mohair production in Southern Africa is confined mainly to S.A., with small contributions from Namibia and Lesotho.

The number of Merino sheep has declined since 1955, with an increase in other woolly sheep, mainly dual-purpose and mutton breeds which offer an income from carcasses. Fashion changes caused a decline in Karakul sheep numbers. Karakul pelts are produced in the arid semi-desert regions of western S.A. and Namibia. The Karakul pelts industry is hampered by a market plagued by price and fashion fluctuations and recently pressure by animal welfare groups too.

The viability of Karakul pelt market depends on the proper integration of the production of quality pelts and manufacturing and marking in a highly competitive world market. Karakul wool is a by-product of the Karakul pelt industry and constitutes some ¾ % of the total income from pelts and wool. Since karakul wool is coarse and colored, it is classed as carpet wool for trade purposes. Karakul wool blankets have been produced in RSA for many years.

If Karakul wool in Namibia is not collected by a local co-operative or sold locally, it must be sent by rail to one of the following addresses: BKB Siding 3074, Port Elizabeth, or SA Wool Board, Siding 512, Isipingo. Bales are weighed strictly in accordance with the information given on the bales.

Advance payment prices for each type are announced by the SA Board at the beginning of the wool season. Assessment of wool for advance payments purposes is done as follows:

- The trading type of the wool is determined with due consideration for the type, colors, length and grass-seed.

- Clean yield – In raw condition wool contains several impurities such as dust, lanolin, sand and grass-seed.

After the advance payment has been done, the wool is classed in the various trading types by the sorters. At the end of the season the difference between the overall advance payment and the net income earned by the Wool Board is paid to the producer by way of a payment in relation to his advance payment.

The value of wool fleece is largely determined by the amount of clean wool that is produces. The length, density and diameter of the wool fibre also affect the value of the wool by determining its grade.

5.4.4 ECONOMICALLY IMPORTANT TRAITS OF WOOL

The commercial value of wool is determined to a large extent by its textile properties. Quality traits are important in the breeding programmes and management. The fibre diameter of RSA wool clips ranges between 17 and 33 micron, 89 % of the clips varies between 20 and 30 micron. Good quality wool is soft and relatively under crimped and has a low resistance to compression.

The most important raw wool characteristics for processing are fibre diameter, staple length, crimp frequency, style index and mean staple strength. Spinning performance is related to fibre diameter of both wool and mohair, with finer fibres spinning to finer yarns. Finer wool shows fibre breakage during processing and produces more oil, partly as a result of a tendency to increased felting. Variation in wool production, quantitatively and qualitatively, is influenced by variety of environmental factors.

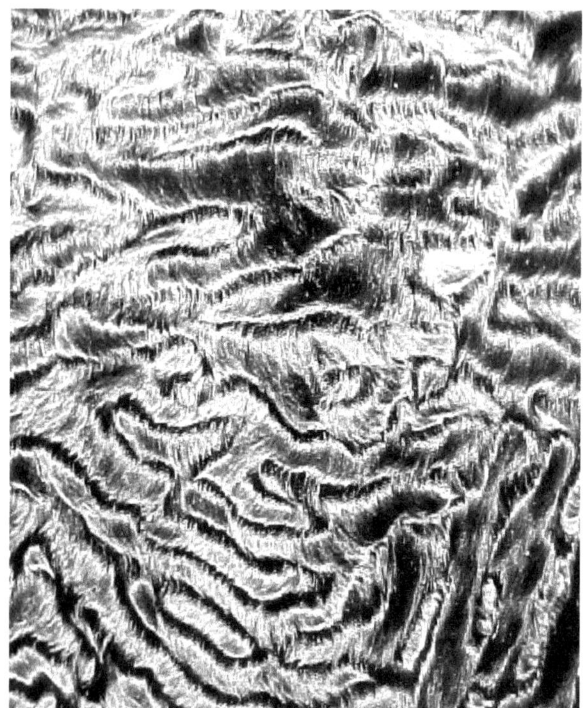

Fig. 5.4 Black Karakul pelt

The average greasy fleece in R.S.A is 5,2 kg and clean yield 3,9 kg. Several breeders are of the opinion that the genetic potential of the Merino and Angora are not fully expressed on available natural pastures. The genetic potential could be realized either by supplementing natural pastures or by artificial pastures.

5.5 HEALTH MANAGEMENT OF SMALL STOCK

5.5.1 INTERNAL PARASITES

5.5.2 INTRODUCTION

Farmers often do not realize to what extent internal parasites can adversely affect their animals and by the time that clinical signs are seen, it is too late.

Mortality rates in small stock can be quite devastating, particularly in hot and humid environments and high population densities.

The external appearance is not always indicative of in-/external parasites infestation. It is usually the growth rate, production and general resistance against disease that suffer most in affected animals.
While the control of diseases and parasites is affected by way of planning programmes of vaccination, de-worming and dipping, management must constantly be focused on steps to avoid or minimize stressful conditions that predispose to infectious diseases and parasitic infestations such as:

- the introduction of animals from disease-free areas to areas where parasitic and infectious problems are rife.
- malnutrition, feeding errors, sudden cold, heat or humidity, transport, confinement.
- sudden and complete changes in veld type, diet or water quality.
- procedures during shearing, weaning, docking and castration.
- poor mothering ability and the exposure of underfed lambs and kids to extremes of cold or heat.

5.5.3 INTERNAL PARASITES CAN BE DIVIDED INTO THREE GROUPS

1) Trematoda - flukes (Liver fluke, Conical fluke)
2) Cestoda - tapeworms (Milkworm, liver tapeworm)
3) Namatoda - roundworms (Wire worm, Bankrupt worm)

The worms all lay eggs which pass out in the faeces for further development if there is sufficient moisture and the temperature is high enough. The larvae, after hatching from the eggs, are ingested by the animals while grazing. The larvae attach themselves to blades of grass near the soil where they are protected against desiccation and irradiation by the sun.

5.5.4 ROUNDWORMS

Roundworms are elongated, cylindrical worms, tapering at each end. Some are very large (*Ascaris* species in calves) while others are barely visible to the naked eye (*Trichostrongylus* species in cattle and sheep / goats). Some roundworm are white in colour while others are reddish-brown or brown.

5.5.5 LIFE CYCLE

Most roundworm species have a direct life cycle, which does not involve an intermediate host. Adult female worm present in the host animal laid eggs, which are passed in the faeces and thus contaminate the veld or pasture.

Heat and oxygen are needed for roundworm eggs to hatch, the eggs will be killed by excessively high temperatures, but not by freezing.
The first and second stage larvae must have moisture in order to survive.

They die if conditions are dry after they have hatched, whereas the third stage or infective larvae , retain their sheaths and therefore they are protected. When warm, moist conditions prevail, the infective larvae climb up the blades of grass and are swallowed by grazing animals. This happens early in the morning or at dusk or when the sunlight is not too bright.

In the dry periods, or when the sun shines brightly, the larvae move downwards and hide near the ground. This is one of the reasons why sheep, being short grazers, are more easily infected than cattle.

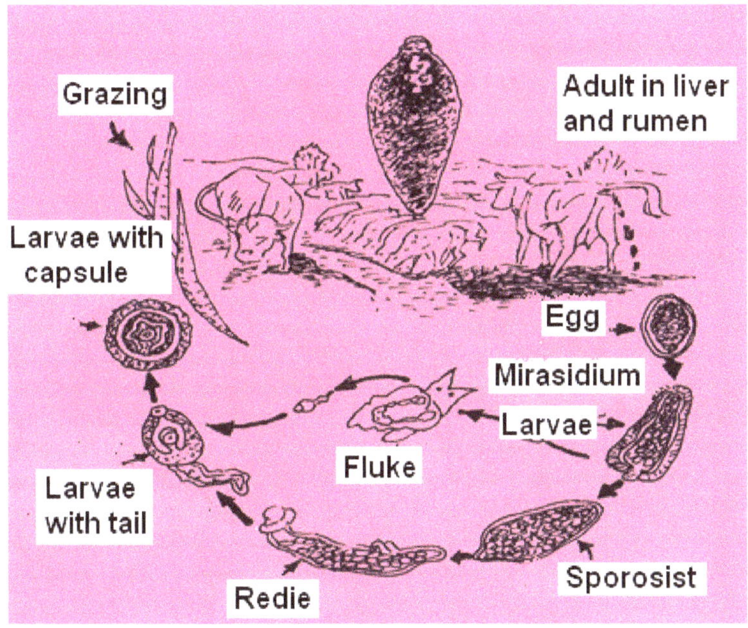

Fig. 5.3 Life cycle of a fluke

In the case of lungworm, the larvae, once swallowed, migrate through the intestinal wall into the blood stream and are carried to the lungs, where larvae development proceeds.

5.5.6 IMMUNITY

With a few exceptions, cattle tend to be resistant to the roundworms of sheep and goats and *vice versa.* Sheep and goats are susceptible to the same roundworm parasites.

Cattle, sheep and goats will acquire a degree of immunity to internal parasites. Young stock, including weaned animals are more susceptible to internal parasites than are adult animals. Poor nutrition during drought or winter periods will increase susceptibility to infestation.

TABLE 5.1 SOME OF THE MORE IMPORTANT ROUNDWORM PARASITES OF STOCK AND WHERE THEY ARE FOUND IN THEIR DOMESTIC HOST SPECIES

SITE	COMMON NAMES	HOST
Abomasums	Wireworm	Cattle, sheep and goats
	Bankrupt worm	Cattle, sheep and goats
Small intestine	Bankrupt worm	Cattle, sheep and goats
	Cattle bankrupt worm	Cattle
	Long-necked bankrupt worm	Sheep and goats
	White bankrupt worm	Cattle, sheep and goats
	Sandveld hookworm	Cattle, sheep and goats
	Cattle and Grassveld hook Worm	Cattle, sheep and goats
	Ascaris	Calves
Large intestine	Nodular worm	Sheep and goats
	Large mouth bowel worm	Cattle, sheep and goats
Caecum	Whipworm	Cattle, sheep and goats
Lungs	Lungworm	Sheep and goats
Sub cutis	Parafilaris	Cattle

5.5.7 CLINICAL SIGNS

Fig. 5.6 Bottle jaw (accumulation of fluid under skin below the jaw)

The visible signs of roundworm infestation are thus any or more of the Following:

- Listlessness
- Loss of appetite
- Lost of weight
- Dry, staring hair coat
- Lagging behind the flock
- Diarrhoea, dehydration, sunken eyes
- Bottle jaw (accumulation of fluid under skin below the jaw).

- Severely affected animals are weak and lag behind the flock. They are listless and sluggish to feed.

- The mucous membranes of eyes, mouth, vulva, etc. are pale or even white due to anaemia.

By the time these clinical symptoms are evident, the economic losses due to decreasing growth and production will have occurred already.

5.5.8 CONTROL OF ROUNDWORMS

There are three basic tactical approaches to roundworm parasite control:

- Veld / pasture management
- Resistance and immunity
- Vaccines / Dosing (worm remedies)
- ***See Fig. 5. 7 Wireworm (Page 97)***
- ***See Fig. 5. 8 Milk tapeworm (Page 97)***

5.5.9 DIAGNOSING WORM INFESTATION BY WORM EGG COUNT

The purpose of this is to determine which nematodes are important to farmers and when they cause problems.

1) Take equal sample of faeces from the rectums of a particular group of animals, calves, sheep and goats.
2) Keep them cool until the examination is done, worm eggs hatch if it is too warm and they burst when they are frozen.
3) Meticulous marking of the samples is essential, must include name of owner, farm name and number, description of animals, age or sex.
4) Send the samples to the laboratory expeditiously where a worm egg count will be done by means of concentration techniques.

5.5.10 VELD / PASTURE MANAGEMENT

Rotational grazing – if stock is removed from veld or pasture for a sufficient period of time, larvae, which hatch, are unable to find a host and will die. Stock must be absent from the veld or pasture for 2 to 3 months. If possible, use should be made of veld grazing at times when pasture infestation is high (e.g. Nov. to Jan. in summer rainfall areas.) If browse constitutes a significant portion of the available grazing, uptake of larvae will be limited. Larvae must crawl up grass stems to be swallowed by their host but are not able to reach browse height.

5.5.11 RESISTANCE AND IMMUNITY

Cattle, sheep and goats will acquire a degree of immunity to internal parasites providing that they are exposed to the particular parasites at a moderate level. Cattle develop a high level of immunity to wireworm, in sheep immunity to wireworm is rare and seldom permanent , but some degree of resistance to wireworm can develop. Poor nutrition during drought periods or disease, lambing, shearing, etc. will increase susceptibility to infestation. When ewes are about to lamb, there is a drop in the level of immunity. Ewes must therefore be dosed shortly prior to lambing and should preferably lamb in a "safe" place. The infestation of ewes and lambs must be monitored until weaning and the lambs must be weaned onto "safe" pasture.

1.5.12 DOSING

Use anthelmintics with an AAA rating for the specific worms obtained from list of anthelmintics; also mentioned on every bottle of anthelmintics. Rotate preparations of the five groups, namely Benzimidazol, Avermectine, and Levamisol/Morantel, Organophosphates and Phenol substitutes and Salicylic Anylides.

1.5.13 EXAMPLE OF A DEWORMING PROGRAM

It must be remembered that great variation exists in worm infestation from area to area, time to time and under different management systems and therefore extrapolation and fixed general deworming programs can only be very general guidelines, to be adapted to the specific situation.

1) After the first rains in September, dose with Seponver Plus, which is effective against wire worm, nodal worm and nasal worm, and has a three week's residual effect against reinfestation with wireworm.
2) In November, dose with Seponver again.
3) In February, dose with Ripercol which is effective against wire worm and nodal worm.

4) In March, dose with Flukiver, which in effective against nasal worm, wire worm and has a seven-week residual effect against wireworm.
5) In May, after the first frost, dose with Seponver Plus.
6) Dose lambs, at the age of 6 weeks, with Multispec which is effective against tapeworms, nodular worm and wireworm. At 12 weeks, dose with Lintex which is effective against tape worm infection, and with Seponver Plus against wire worm, nodular worm and nasal.

5.5.14 VACCINES / DOSING

Experimental vaccines have been developed and show promising results for internal and external parasites. Dosing remedies for control of internal parasites have been in use for more then 20 years, however, they must be used intelligently if parasites resistance to the active ingredient is not to become a serious limiting factor. The product selected will depend on the parasite or parasites prevalent at that time and whether any parasite resistance is known or suspected. A narrow-or broad- spectrum remedy may be selected accordingly. It is essential to alternate between the different chemical groups in order to prevent internal parasites from developing resistance to the chemical groups.

Stock should be given the correct dose, according to their body mass. Under dosing encourages the development of resistance, overdosing wastes money and may even result in stock losses due to toxicity. Check the expire date, if applicable, and shake the mix remedies unless recommended otherwise by the manufacturer. The frequency of dosing will depend on a number of factors, including whether management, extensive or intensive pasture, type and age of the stock etc. The efficiency of the dosing program can be monitored by means of regular routine faecal round egg count.

5.5.15 EXTERNAL PARASITES CAN BE DIVIDED INTO THREE GROUPS

According to their life cycle, ticks can be divided into three groups:

1) One-host ticks: here all three stages (larva, nymph and adult) stay on one animal and do not leave it to moult.
2) Two-host ticks: the larva and nymph stays on one animals and the adult on another animal.
3) Three-host ticks: each stage parasitises a different host, most ticks belong to this group.

The female lays a large number of eggs (5000 – 10 000 or more) in a secluded spot, larvae hatch from eggs (larvae are six-legged), they engorge on blood. The moult and the nymphs (with eight legs) appears, the nymphs engorge themselves on blood, moult and the adult male and female ticks appear. They again feed on blood, mate and the cycle repeats itself.

Ticks have a direct detrimental effect on the animal in that:

1) They damage the skin.
2) They lower the resistance of the host as well as cause condition loss
3) The bite wounds are painful, especially in imported animals, not accustomed to S.A. ticks, swelling may occur.
4) Wounds, caused by the long mouthpart of some species, present ports of entry from secondary organisms to enter the body.
5) Their presence results in the loss of teats, ears and or the types of tails and cause abscesses and lameness.
6) Anaemia, as a result of the large volume of blood sucked by them, often occurs.
7) Deaths as a result of large numbers of ticks can occur.

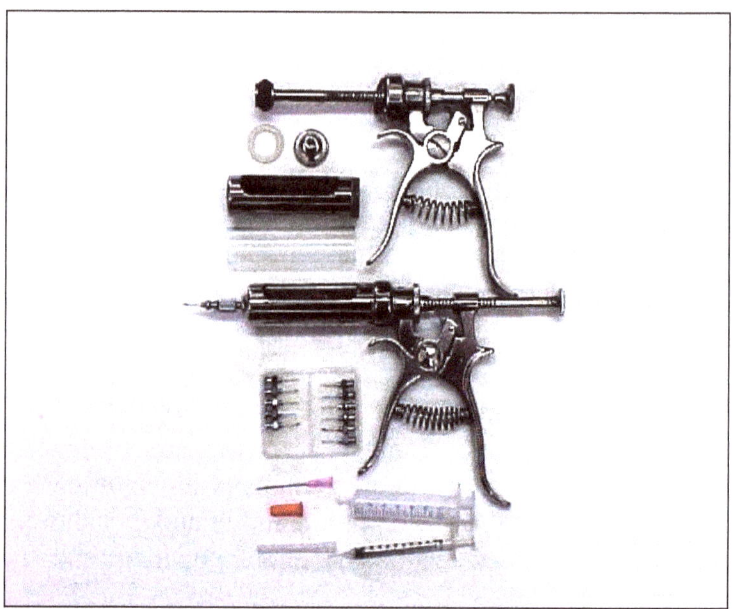

Fig. 5.9 Roux Syringe and needles, Disposable Syringe and needles.

5.5.16 THE CONTROL OF TICKS, LICE AND MITES

As ticks annually cause damage worth millions of rands every year, stock owner are forced to take some control measures to limit their numbers:

1) Burning of the veld will definitely reduce their numbers but only for short periods. Only those on top of the soil will be killed and not those hiding away in some sort of shelter.
2) Rotational grazing, actually serves little purpose since some stages can go without food for very long periods of time.

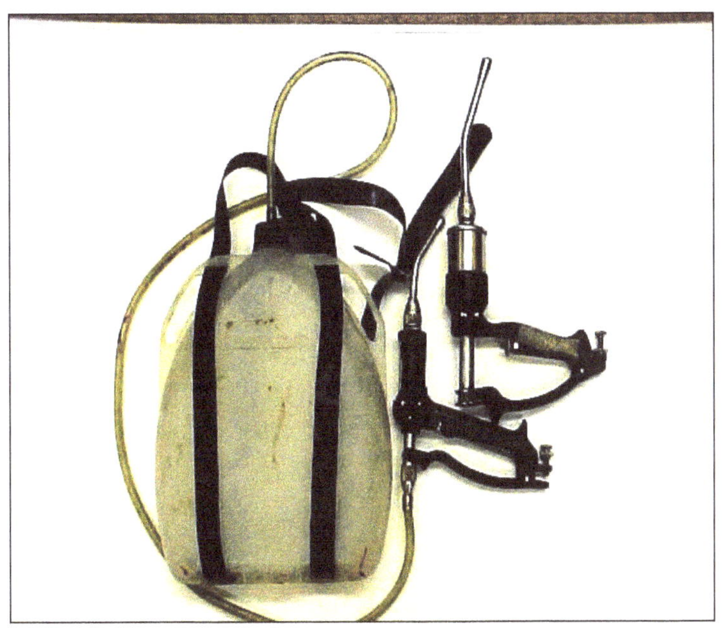

Fig. 7.10 Auto Dosing Syringe.

3) The most practical method still remains dipping (Zip dip, Dazzel, Cooperzon dip and Paracide dip) or spraying the host in/with approved and safe remedies (Delete-, Sypor- and Clout pour on). If this is being done regularly the occurrence of diseases and the number of ticks will be limited to a minimum. The dipping compound should always be mixed according to the instructions on the container. Animals can also be inject with Ivomac or Dectomac (ml / 50 kg) under the skin, which kills internal / external parasites.

5.6 DISEASES

Small stock should as far as possible be vaccinated eight weeks prior to mating. The following are, however diseases that should receive attention when farming with small stock.

5.7 FOOT ROT

5.7.1 CAUSE AND DISTIRBUTION

Foot rot is caused by Bactericides nodosus. The organisms widely distributed in nature, particularly under wet muddy or moist conditions, enter the body through small wounds or waterlogged hooves usually at the coronary band.

5.7.2 CLINICAL SIGNS AND POST MORTEM LESIONS

Foot rot affects small stock at all ages with young lambs being most susceptible. The conditions is very painful, the animal limps badly and the affected area is very hot. The foot rot hoof is often lost.

5.7.3 DIAGNOSIS, TREATMENT AND PREVENTION

The diagnosis of the conditions is not difficult. Treatment requires time and effort and should be tackled locally as well as systemically. During wet conditions muddy kraals, marsh camps and cultivated pastures should be avoided, by moving animals to higher well-drained camps. Outbreaks can also be controlled by driving animals through a shallow footbath containing 10% blue stone (copper sulphate) or zinc sulphate solution at least once a week when wet conditions prevail.

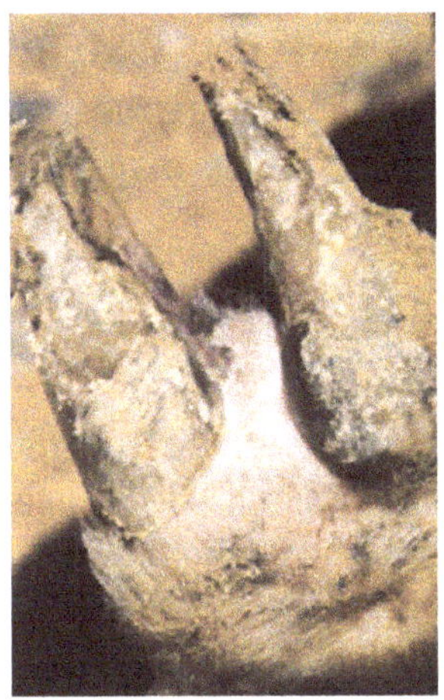

Fig. 5.11 Foot rot, small stock.

5.8 RIFT VALLEY FEVER

5.8.1 CAUSE AND DISTRIBUTION

Rift valley fever is an infectious insect-born, viral disease of sheep, goats and cattle. Man is also susceptible to Rift valley fever. A virus, which is transmitted, causes rift valley fever to susceptible sheep by mosquitoes. Mosquitoes normally breed in stagnant water, especially in low-lying areas such as veils, rivers, dams and pans. Hot weather and good summer and autumn rains are conducive to breeding of large numbers of mosquitoes, which are then available to transmit and spread the disease. In the case of human beings, it is transmitted during handling of infected carcasses (*post mortem*). Keep livestock away from low-lying areas such as rivers, dams and pans during summer times, because mosquitoes normally breed in stagnant water.

5.8.2 CLINICAL SIGNS AND POST MORTEM LESIONS

Sheep, and especially young lambs, are most susceptible. Up to 95% or more of any young group of lambs may die within a relatively short period.

Usually the first sign of the disease to be observed is an increasing daily death rate. Some lambs may die within 12 to 24 hours of the onset of the disease without showing any clinical signs. Other lambs may linger on for days showing clinical signs such as fever, weakness and often bloody diarrhoea.

Adult sheep show fever, loss of appetite, bloody diarrhoea and trembling weakness. Approximately 30% of affected sheep usually die within 2 to 7 days of the onset of the disease. Pregnant ewes, which contract the disease often abort promptly particularly during rainy seasons and late summer, but sometimes a variable percentage will give birth to premature, weak and listless lambs, which usually die within a day or two.

5.8.3 DIAGNOSIS, TREATMENT AND PREVENTION

Post Mortem – The liver is enlarged, swollen and friable, and brown to yellow in colour with small grayish-white spots over the surface and throughout its substance. The spleen is swollen and there is a reddish-brown fluid in the abdominal and chest cavities as well as the heart-sac.

The lining of the intestine may be red and the contents are often bloody. Rift valley fever and Wesselsbron disease are often confused, it is advisable to consult the nearest veterinarian. Treatment is of no avail. Control of this disease is based mainly on preventive vaccination. An effective live vaccine for small stock is available. This vaccine usually affords lifelong immunity after a single injection.

It will only be necessary to vaccinate the lambs annually. Small stock should be vaccinated stock at any age, except lambs born from ewes that have already been vaccinated. These lambs should not be vaccinated before they are 5 to 6 month old. Such lambs acquire a passive immunity from their mothers. Do not vaccinate pregnant ewes. Keep livestock away from low-lying areas such as rivers, dams and pans during summer times, because mosquitoes normally breed in stagnant water.

See Fig. 5.12 The lining of the intestine may be red and the contents are often bloody. (Page 98)

5.9 WESSELSBRON

5.9.1 CAUSE AND DISTRIBUTION

Wesselsbron is an infectious insect-born, viral disease of sheep, goats and cattle. Man is also susceptible to Wesselsbron. A virus, which is transmitted, causes Wesselsbron to susceptible sheep by mosquitoes.
In the case of human being, it is transmitted during handling of infected carcasses (*post mortem*). Keep livestock away from low-lying areas such as rivers, dams and pans during summer times, because mosquitoes normally breed in stagnant water.

5.9.2 CLINICAL SIGNS AND POST MORTEM LESIONS

For all practical purpose, Wesselsbron disease can be regarded as an abortion disease, since adult small stock, apart from a possible mild fever reaction, do not normally show visible clinical signs of the disease.

The most important manifestation of this disease is that a large percentage of pregnant ewes either abort or deliver still born lambs without being affected any other way. Some lambs may be born alive, but are weak and usually die with in 24 hours. Apparently normal lambs can become infected shortly after birth and within a few days up to 30 % die. As in the Rift valley fever, the virus primarily affects the liver.

5.9.3 DIAGNOSIS, TREATMENT AND PREVENTION

Post mortem – As in the case of Rift valley fever, the virus primarily affects the liver. The liver will be swollen and light yellow to yellowish –brown and very friable. The carcass of young lambs are usually very pale, with a bloody fluid in the abdominal and chest cavities. The intestinal canal is usually red and the contents often bloody. Adult small stock may also show a swollen, yellow and friable liver. Wesselsbron and Rift valley fever disease are often confused, it is advisable to consult the nearest veterinarian. Treatment is to no avail. Control of this disease is based mainly on preventative vaccination. An effective live vaccine for small stock is available. This vaccine usually affords lifelong immunity after a single injection. Small stock should be vaccinated when they are 5 to 6 month old, mature small stock at any age. Do not vaccinate pregnant ewes. Keep livestock away from low-lying areas such as rivers, dams and pans during summer times.

See Fig. 5.13 The liver will be swollen and light yellow to yellowish brown and very friable. (Page 98)

Fig. 5.7 Wireworm

Fig. 5.8 Milk tapeworm

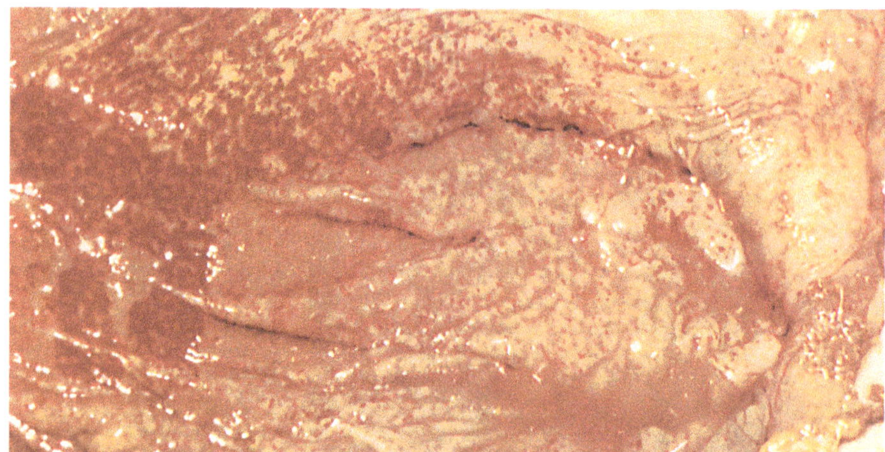

Fig. 5.12 The lining of the intestine may be red and the contents are often bloody.

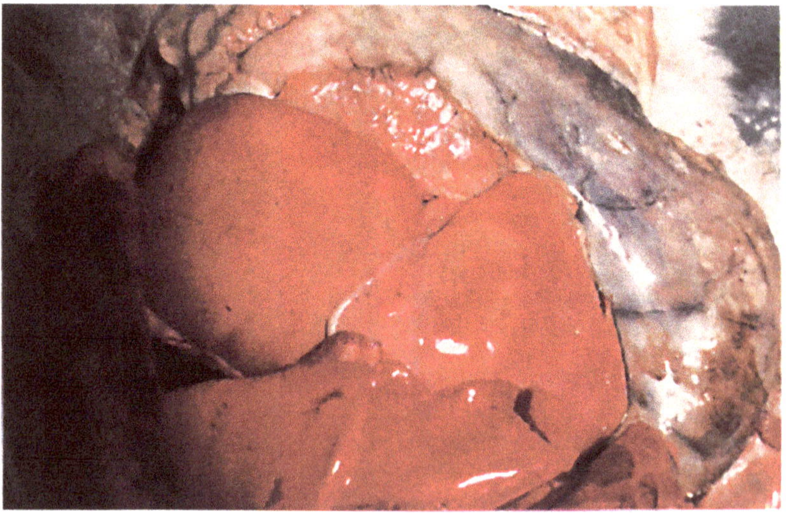

Fig. 5.13 The liver will be swollen and light yellow to yellowish brown and very friable.

5.10 BLUE TONGUE

5.10.1 CAUSE AND DISTRIBUTION

Blue tongue is an infectious insect-born, viral disease of sheep. A virus, (of which 21 different strains are presently known) which is transmitted by midges causes Blue tongue to susceptible sheep. Because it is spread by midges, bluetongue has a typical seasonal incidence, normally occurring from mid-summer to autumn, or from the first good rains until the first frost.

When good rains occur or where pools or veils still contain water as a result of ample rain the previous season, and if hot weather conditions prevail, bluetongue can make an earlier appearance. Keep livestock away from low-lying areas such as rivers, dams and pans during summer times, because midges normally breed in stagnant water.

5.10.2 CLINICAL SIGNS AND POST MORTEM LESIONS

Because the various strains of the blue tongue virus differ in their virulence, the severity of the disease and noticeable clinical signs will also vary according to the involved. Other important factors in this respect are the breed, age and conditions of the sheep, and whether and how often the sheep were vaccinated previously. Most of the various strains are included in the blue tongue vaccines, but resistance cannot be build up by the sheep after a single injection. The first clinical signs to be noticed are fever, listlessness, lost of appetite and rapid breathing. The mucous membranes of the nose, mouth and eyes turn red and the lips become swollen. Eventually the mucous membranes of the mouth and tongue turn purplish-blue in colour and develop small ulcers, especially on those parts in contact with the teeth. Initially there is a watery discharge from the nose which later turn grayish-white to yellow.

In the acute stage of the disease, the animals appear stiff or lame and may show a tendency to walk on their knees or lie down. Although the stiff and painful gait is noticeable, the typical signs of inflammation of the coronets only become visible 4 to 5 days later. A reddening which appears at the junction between the skin and hoof and gradually moves downward. The red colour darkens and is followed by the formation of a distinct ridge or crack in the hoof.

See Fig. 5.14 Eventually the mucous membranes of the mouth and tongue turn purplish-blue in colour and develop small ulcers, especially on those parts in contact with the teeth.(Page 103)

5.10.3 DIAGNOSIS, TREATMENT AND PREVENTION

Post mortem – apart from the clinical signs described, the intestines are often reddened and the spleen enlarged, while haemorrhages are also found on the inner and outer surfaces of the heart, and usually also in the wall of the pulmonary artery.

There is no specific treatment for blue tongue. Treatment is therefore aimed at relieving the symptoms and proper nursing by providing shade, shelter, drinking water and easily accessible food. Antibiotic treatment is given to prevent secondary bacterial infection. Sick sheep should never be forced to move. Regular preventive vaccination remains the best method of control of the disease. The available vaccines afford protection against most strains of the virus and should be given annually. The best period for vaccination is from Aug. to Dec. With spring mating and autumn lambing no problem should be encountered, since vaccination can be conducted in the pre-mating period. Rams must be vaccinated 8 weeks before the breeding seasons and ewes 9 weeks before mating. Lambs from unvaccinated ewes should be vaccinated at the age of 2-4 months.

5.11 CORYNEBACTERIUM

5.11.1 CAUSE AND DISTRBUTION

This disease is normally characterized by abscessation of the lymph glands but can also manifest itself as individual abscessation in the body, scrotum, or internal organs, especially the lungs.

The organism can survive in manure for long periods with resultant buildup of infection in kraals. When operations are carried out in these kraals the wounds may come in contact with the infection and the bacteria gain entrance into the body. However, the shearing shed, infected clothes of shearers and their shears constitute the most important means of transmission. These abscesses are often accidentally cut during shearing or are purposely lanced by the shearers, using their shears.

5.11.2 CLINICAL SIGNS AND POST MORTEM LESIONS

The most common form of this infection is characterized by the formation of abscesses in the lymph glands in front of the shoulder, in the hind leg and in the loin immediately in front of the hind leg. The abscesses develop slowly and can be felt and often seen. Initially they are firm or hard and become soft later.

See Fig. 5.15 Abscesses can also be found between the lungs. (Page 103)

Abscesses can also be found in the lymph glands below the ears, the lower jaw as well the sides of the neck in goats, and between the lungs. In the latter cases, the abscesses rupture into the thoracic cavity, forming a purulent inflammation ('harsslagsiekte'/ pluck disease) which, in milder cases causes adhesions between the lungs and chest wall. Such animals may show rapid, but more often chronic, emaciation, laboured breathing and tire easily when driven.

5.11.3 DIAGNOSIS, TREATMENT AND PREVENTION

Treatment usually consists of the draining of the abscesses and treating the cavity with a bactericidal agent. Abscesses are treated locally while systemic infections are treated with antibiotics if still in the acute stage and not too extensive. Prevention is mainly based on proper hygiene measures in the shearing shed. A vaccine is available, and sheep should be vaccinated from the age of two weeks and older. Animals vaccinated for the first time should receive two injections with an interval of six weeks. Immunity is also of short duration (5 months).

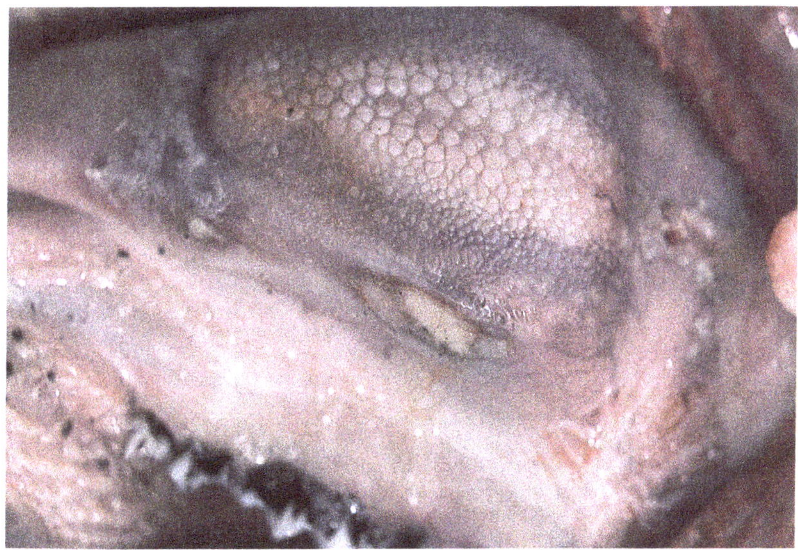

Fig. 5.14 Eventually the mucous membranes of the mouth and tongue turn purplish-blue in colour and develop small ulcers, especially on those parts in contact with the teeth.

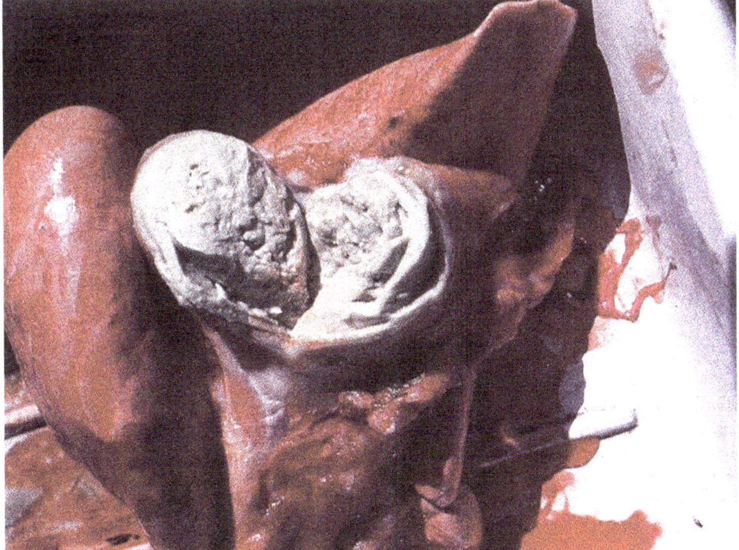

Fig. 5.15 Abscesses can also be found between the lungs.

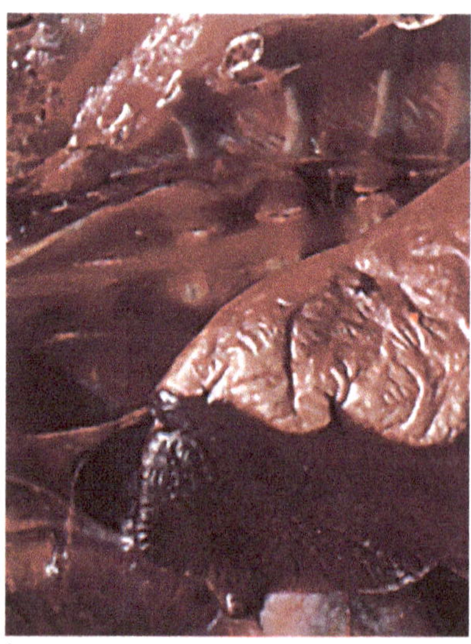

Fig. 5.16 Septicaemia: all organs are bloody in appearance.

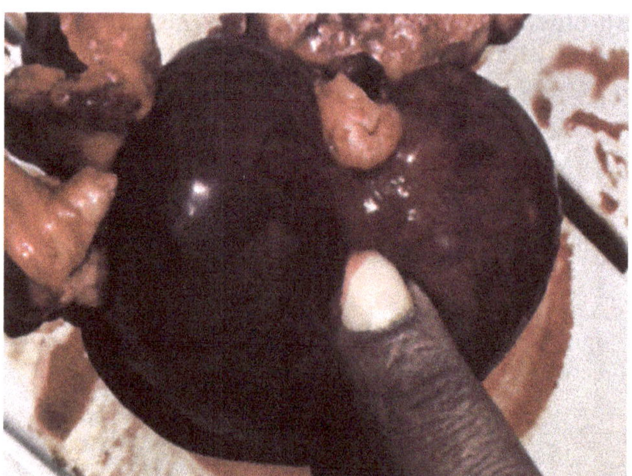

Fig. 5.18 Pulpy kidney: the kidneys, in a fresh carcass, are swollen and dark-red.

5.12 PASTEURELLOSIS

5.12.1 CAUSE AND DISTRIBUTION

This disease is normally characterized by pneumonia or septicaemia and usually occurs in animals subjected to stress conditions, caused by e.g. transporting of animals, change of diet, or sudden change in climatic conditions. Animals in a poor condition are more susceptible.

5.12.2 CLINICAL SIGNS AND POST MORTEM LESIONS

General symptoms of septicaemia such as a fever, increased rate of respiration, coughing and poor appetite. Young lambs are most susceptible and high percentages of mortalities may be experienced. At post mortem varying degrees of pneumonia and pleurisy will be found. The lungs show smaller or larger areas of consolidation which vary in colour from dark red to grey. Haemorrhages are found on the inner surfaces of the heart while the mucous lining of the nasal passages is dark in red colour. The conditions caused by these organisms include pneumonia, inflammation of the liver and mastitis.

See Fig. 5.16 Septicaemia: all organs are bloody in appearance.(Page 104)

5.12.3 DIAGNOSIS, TREATMENT AND PREVENTION

It is essential to consult a veterinarian to obtain an accurate diagnosis by laboratory examination, and treat affected animals as soon as possible. Isolation of the organism from the infected tissues. Pneumonia, with the possible occurrence of abscesses and in the case of septicaemia all organs are bloody in appearance. Early treatment with broad-spectrum antibiotics may be successful. Elimination of stress conditions. Animals vaccinated for the first time should be inoculated twice with an interval of 4 weeks followed by a single injection every six months. Small stock can be vaccinated from the age of two weeks. Pasteurellosis vaccine is effective.

5.13 RINGWORM

5.13.1 CAUSE AND DISTRIBUTION

A skin condition caused by a fungus.

5.13.2 CLINICAL SIGNS AND POST MORTEM LESIONS

Round or hairless patches occur on any part of the skin, especially on the head and neck. Flaky lesions sometimes form around the eyes.

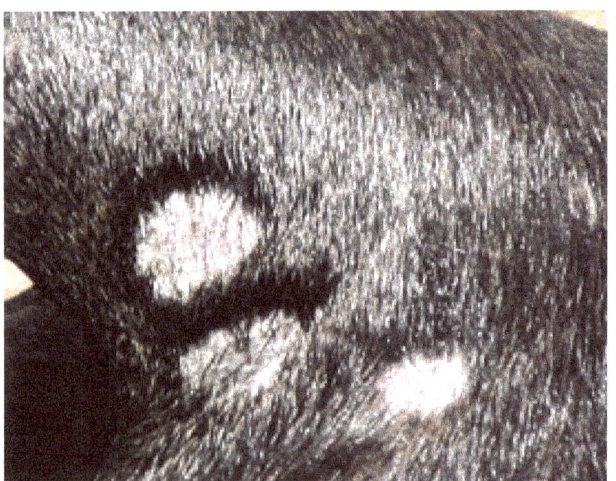

Fig. 5.17 Ringworm

5.13.3 DIAGNOSIS, TREATMENT AND PREVENTION

Skin scrapings taken from the edge of the lesion can be send to veterinary laboratory for confirmation.

All livestock can be dipped with lime sulphate or Imaverol dip, or hairlcss patches treated with ringworm ointments.

5.14 PULPY KIDNEY

5.14.1 CAUSE AND DISTRIBUTION

A toxin produced by the organism Clostridium causes this disease welchii type D.

The predisposing factor usually is some form of digestive disturbance, brought about by various circumstance e.g.

a. The sudden availability of large quantities of good quality food after a period of starvation.
b. Sudden changes from poor to good grazing.
c. Sudden changes from natural to cultivated pastures.
d. The provision of supplementary drought feeding or high concentrates diets with insufficient roughage.
e. Fattening rations without proper adaptation.

5.14.2 CLINICAL SIGNS AND POST MORTEM LESIONS

Pulpy kidney is characterized by the occurrence of sudden deaths and the fact that sheep seldom, if ever, show symptoms before death. A typical feature is that a few sheep will be missed and found dead each morning in a particular camp.

In affected animal showing symptoms two main forms of the disease can be distinguished on the symptoms, namely:

(1) The depressed or listless form.
The affected animal lag behind the flock, tire easily and have a staggering uncertain gait. The animal stumbles easily or knuckles over at the fetlocks, lies down in the upright position with the head turned to one side or lies flat on its side, and die shortly afterwards

(2) The nervous form.
This form is characterized by restlessness, hypersensitivity, staggering, chewing movements and grinding on teeth and irregular breathing. Such animals go down with muscular spasms and kicking movements. The head is often drawn backwards and sometimes there is froth at the mouth. They often leap into the air, fall down and die. On post mortem examinations there are few typical lesions. The carcass is usually bloated with froth present at the mouth, nose and in the trachea. There are small haemorrhages in the skin, especially those parts covering the neck, withers and back. The kidneys, in a fresh carcass, are swollen and dark-red with congestion of the superficial blood vessels. The carcass, especially during hot weather, decomposes rapidly. As the kidneys are most prone to decomposition they become pulpy very rapidly. The urine, if tested for sugar, is usually positive.

5.14.3 DIAGNOSIS, TREATMENT AND PREVENTION

When the urine test for sugar is positive, pulpy kidney can almost certainly be accepted as being the cause of death. It is important that the nearest veterinarian be consulted for an early, accurate diagnosis. There is no treatment for pulpy kidney thus prevention is of utmost importance.

At present there are two vaccines available against pulpy kidney. The one is an oil based vaccine while the other one is an alum- precipitated vaccine.

The alum-precipitated vaccine is administered twice, with an interval of 4 to 6 weeks, again 3 months later and then annually. Young lambs are usually vaccinated for the first time at weaning. When pregnant ewes are vaccinated their young lamb will acquire a passive immunity through the colostrum, the protection is for approximately 6 months. With outbreaks, especially if they have not been vaccinated before, vaccination should be carried out immediately.

See Fig. 5.18 The kidneys, in a fresh carcass, are swollen and dark-red. (Page 104)

5.15 ENZOOTIC ABORTION (CHLAMYDIOSIS)

5.15.1 CAUSE AND DISTRIBUTION

This infection is caused by one of the many strains of chlamydeous psittaci.

Although little is know about the precise method of transmission, experimental observations indicate that infection is picked up by mouth and that afterbirth, uterine discharges and dead lambs, constitute the most important source of infection.

5.15.2 CLINICAL SIGNS AND POST MORTEM LESIONS

When the disease appears in a flock for the first time, up to 80 % of the pregnant ewes may abort. The affected ewe does not show any signs of disease or pending abortion. The symptoms may vary from early embryonic death and resumption (which will only be reflected in poor lambing percentages), early abortions, usually from three months pregnancy onwards and stillborn. Small undersized lambs are born which normally die within two to three days. The post mortem lesions of abortion itself are seldom typical and evidence of the presence of necrotic tissue only, will be found in the cotyledons.

In case of abortions or premature birth, the afterbirth may be retained and metritis may develop. Lambs that die shortly after birth often show areas of consolidation of the lungs. If lambs are born alive nervous symptoms such as shivering, convulsions, stiff neck, weakness in the hind quarters and even total paralysis may develop.

5.15.3 DIAGNOSIS, TREATMENT AND PREVENTION

The symptoms of this disease are usually fairly typical, but a definite diagnosis can only be made by the demonstration of the organisms on microscopic smears prepared from afterbirth, lungs and other lesions. It is therefore advisable to have suspected cases examined by a veterinarian.

Antibiotics may be administered to affected animals but is usually unsuccessful. An efficient vaccine is available. During outbreaks all animals should be vaccinated. Normal vaccination takes place annually. Ewes must be vaccinated four weeks before mating season starts.

5.16 ORF (VUILBEK, SCABBY MOUTH)

5.16.1 CAUSE AND DISTRIBUTION

This is a widespread contagious viral disease of sheep and goats. Orf is caused by a virus and affects mainly the lips and areas around the mouth.

Thorny veld or pasture which may injure the lips and mouth promote the spread of the disease. The virus itself is very resistant, and can survive for months in the scabs of the lesions. Orf is primarily a disease of lambs and kids. Once a young lamb or adult sheep has contracted the disease and has recovered, the animal will retain a lifelong immunity.

5.16.2 CLINICAL SIGNS AND POST MORTEM LESIONS

Orf is characterized by the appearance of small pimples or pox-like lesions on the lips and around the mouth and nose. These lesions increase in numbers and size and spread to adjacent areas e.g. ears, around eyes, coronets, hooves and the udder and teats of the ewe and scrotum of the ram. In uncomplicated cases, healing takes place within 8 to 14 days, scabs become dry and drop off with complete healing of the underlying surface. If secondary bacterial infection occurs, however, the lesions become septic and the conditions can become very serious and of long duration.

5.16.3 DIAGNOSIS, TREATMENT AND PREVENTION

There are no typical *post mortem* lesions apart from those described above. There is no specific treatment, but support treatment should be provided to relieve the symptoms and to prevent secondary infection. The affected areas, especially the mouthparts, should be treated with oily substances to keep the lesions soft for eating and drinking. Such remedies should contain antibiotics or sulphate-drugs to prevent secondary bacterial infection. Severe outbreaks of Orf can be successfully controlled by preventative vaccination.

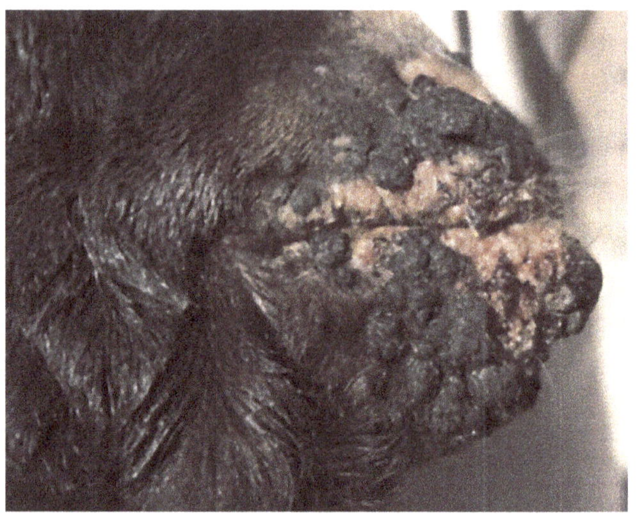

Fig. 5.19 Orf is characterized by the appearance of small pimples or pox-like lesions on the lips and around the mouth and nose.

A vaccine is available from Onderstepoort, or can be also provided by the nearest veterinarian. The vaccine is applied to the inner thigh by dipping the tip of a thigh hypodermic needle into the vaccine and making a few fairly deep scratches into the skin.

5.17 LIVER FLUKE

5.17.1 CAUSE AND DISTRIBUTION

The life cycles of these fluke are all indirect involving an intermediate snail host. The snails are prevalent in areas with permanent water or semi-permanent water supplies. Liver fluke is parasites in the bile ducts of the livers of cattle, sheep and sometimes goats. The mature fluke leaf-like parasites have two suckers for attachment. Fluke eggs pass down the bile ducts and into the intestinal tract of infested animals and are passed in the faeces. The eggs must be freed from the faeces in water in order to hatch, either on the water surface or on blades of grass and then swallowed by livestock.

5.17.2 CLINICAL SIGNS AND POST MORTEM LESIONS

Massive infestations can result in acute death due to haemorrhage caused by the young fluke migrating through liver tissue, less severely affected animals lose condition and become anaemic. Bottle yaw may develop in more chronic cases.

5.17.3 DIAGNOSIS, TREATMENT AND PREVENTION

Post Mortem – There may be fluid in the abdominal cavity, as well as bottle jaw. There will be fibrosis of affected areas of liver tissue, these areas become shrunken, have an irregular surface and are very firm. Fluke can be squeezed out of the bile ducts. Infestation takes place in spring, summer and autumn. Remedies, which are effective against immature and mature fluke can be dosed to prevent clinical signs and reduce pasture infestation. If liver fluke is a serious problem on the particular farm, dams and streams should be fenced off. Water troughs should provide water for livestock.

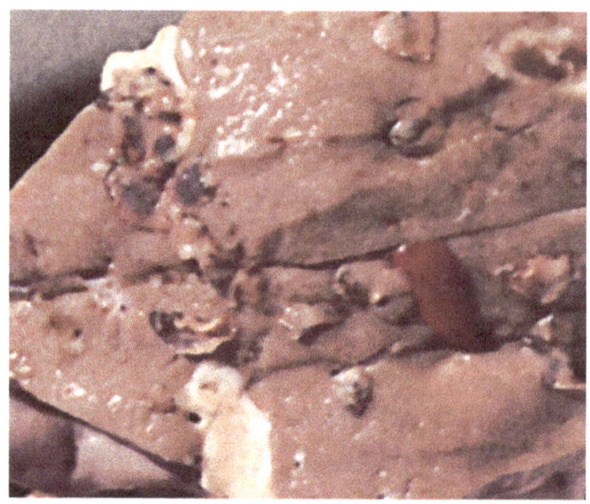

Fig. 5.20 Liver fluke.

5.18 BLUE UDDER (MASTITIS)

5.18.1 CAUSE AND DISTRIBUTION

Various organisms have already been incriminated as the cause of mastitis in small stock via: Pasteurellosis, Corynebacterium. It is not known when these organisms, which occur everywhere, gain entrance into the udder. Although introduction in most cases takes place via the teat canal shortly before symptoms appear, the organism can probably also be harboured in the udder tissue for long periods before inflammation, usually triggered off by bruising of the udder, develops. Mastitis in small stock occurs as a very acute or less acute disease immediately after or within one week of lambing. Pasteurellosis hemolytic is usually involved in these cases, which occur about four to eight weeks after lambing and, although it may be very acute, it is always less acute than Staphylococcus mastitis.

5.18.2 CLINICAL SIGNS AND POST MORTEM LESIONS

The typical acute form is characterized by the sudden swelling of one or both side of the udder. The affected side is hot, painful and initially light to dark red in colour in sheep with a light skin. The udder becomes extremely painful and the condition is accompanied by a high temperature, loss of appetite and rapid breathing. The ewe refuses to allow the lamb to suckle, thus the lamb bleats continuously and appears thin and starved. The red colour of the udder soon changes to a reddish-purple and eventually a blue or greenish-purple colour. In acute cases there are signs of general blood poisoning, the udder has a typical enlarged blue or greenish –purple appearance.

5.18.3 DIAGNOSIS, TREATMENT AND PREVENTION

The diagnosis of the condition usually does not present any problems but as so many organisms may be involved, the specific causative organism can only be identified by a detailed laboratory examination. Treatment consists of specific mastitis remedies administered through the teat canal as well as systemic treatment with antibiotics. In most acute cases, when symptoms are noticed, the condition is usually too advanced for effective treatment. Although treatment might save the animal's life, the entire udder or part of it is permanently lost so that such an ewe has to be culled as unfit for breeding purposes. It is therefore, essential to have specimens examined, not only to determine which organism is involved, but in case of Pasteurellosis to identify the specific strain involved.

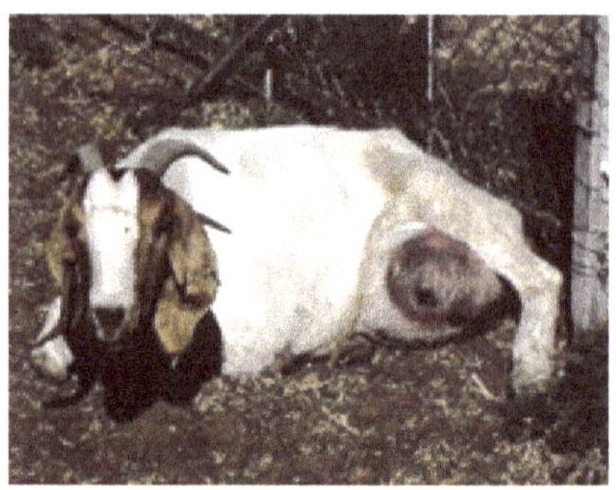
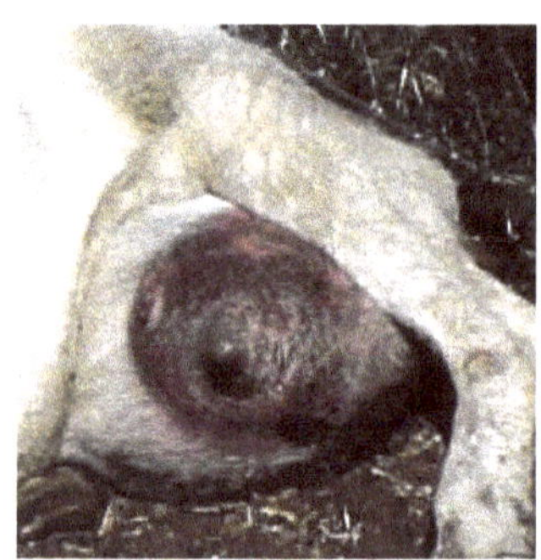

Fig. 5.21 Blue udder, painful and initially light to dark red in colour in sheep with a light skin.

5.19 PROPOSED IMMUNIZATION PROGRAMME FOR SHEEP AND GOATS THAT HAVE NOT BEEN IMMUNIZED BEFORE

This calendar was drawn up for commercial farming and is intended for a farmer who intends having one lambing season per year. Individual farmers may have different views concerning lambing times.

TABEL 5.2 IMMUNIZATION PROGRAMME FOR SHEEP AND GOATS

TIME AND AGE OF ADMINISTRATION	ESSENTIAL VACCINES	
9 - Weeks before breeding season	Bluetongue A,B,C, (3 - weeks) – Ewes	Sheep, Goats
4 to 6 weeks before breeding season	Rift Valley fever, Wesselsbron and Enzootic abortion	Sheep, Goats
Just after breeding season	Bluetongue A,B,C, (3 - weeks) – Rams	Sheep, Goats
6 to 8 weeks before lambing season	Tetanus	Sheep, Goats
From 2 weeks of age	Pasteurellosis, Corynebacterium	Sheep, Goats
Weaning age	Brucellosis Rev I, Pulpy Kidney, Lamb dysentery	Sheep, Goats
6 Months of age	Bluetongue A,B,C, (3 - weeks), Rift Valley fever, Wesselsbron and Anthrax	Sheep, Goats

CHAPTER 6: GRAZING HABITS OF SMALL STOCK

6.1 INTRODUCTION

Herbage production consists essentially of the conversion of solar energy (in the presence of chlorophyl), soil nutrients, air and water into herbage with a certain nutritional value. The genetic constitution of the plant determines not the efficiency of conversion of energy but also the uptake of mineral nutrients, and thus the chemical composition and nutritional value of the herbage. The nutritional value of a pasture, or plant species, can be defined, from the animal viewpoint, as the degree to which the pasture, or species, is able to provide the nutritional needs of the animal in question and is thus the animal response per unit of the intake.

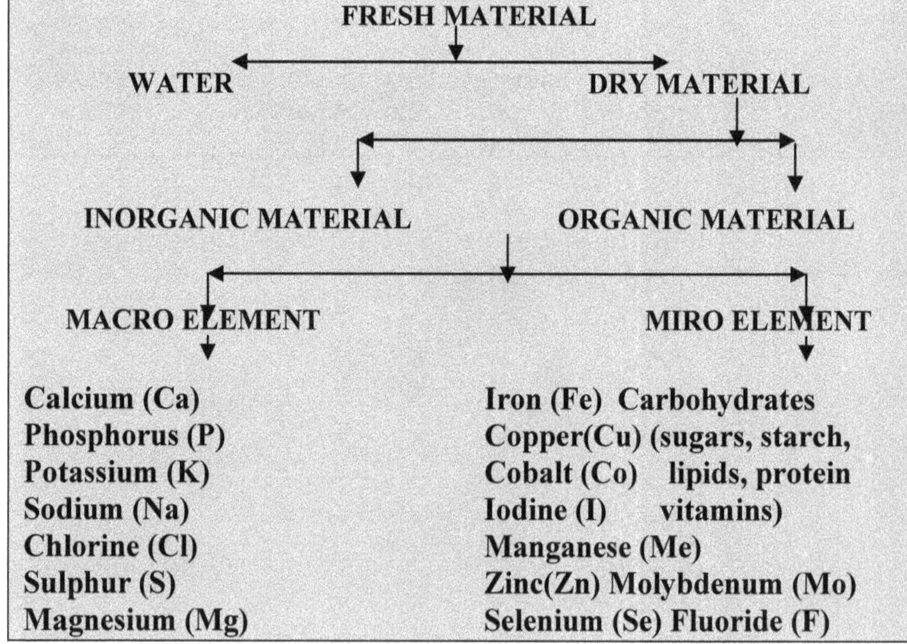

Fig. 6.1 The chemical composition of herbage plants is illustrated.

The grazing habits of small stock and cattle differ widely and are determined by the climate. Difference in grazing habits of different breeds is also encountered (for example Merino and Dorper sheep). In hotter weather the animals will graze for shorter periods during the day and for longer periods during the night. A rise in temperature also lead to an increase in the number of times an animal defecates, urinates and drink water. Cold windy and rainy weather shortens the grazing time. Adopted animals do not react that strongly to climate changes. Management aspects also play a role in the grazing habits of animals. Sheep kept in jackal-proof camps will graze less and when supplementary feeding is stop, they lose condition. Any changes must therefore be introduced gradually.

6.2 THE SENSES

Smell - Animals do not like to graze where other animals have defecated, lush growth near dung-heaps is usually avoided.

Touch - Sheep do not like eating hairy plants whereas goats are not so selective.

Taste – Little is known about the ability of sheep or goats to differentiate between different tastes. The variety of available plant material to a large extent determine how selective the animals will be. Sheep accustomed to grassland will take a long period to adapt to cultivated land.

Fig. 6.2 Dorper sheep

6.3 SHEEP

Although sheep eat a large quantity of grass, the grass must be young and green before they utilize it properly. Sheep are very selective and primarily low grazers. They utilize the vegetation between 2,5 and 20 cm. Sheep utilize both grass and bushes equally well, but concentrate on grass in summer and on bushes in winter. Sheep utilize small shrubs and perennial herbage to a greater degree than cattle, but to a lesser degree than Meat goat. Their split upper lips facilitate selection of parts of plants and short grazing. They are less vigorous eaters than the non-cloven hoof non-ruminant animals such as horses and donkeys and then do less damage. It is accepted that of the different sheep breeds, the Merino is the most selective grazer while the other hardier breeds such as the Dorper and Karakul are less selective but heavier grazers. The damage caused by sheep is due mainly to the fact that they graze so short and selectively. This usually causes detrimental changes in the botanical composition which are followed by other symptoms of overgrazing. Sheep (just like cattle) frequently walk 4 kilometres in 24 hours, but also as little as 0.64 km. In the Karoo it is often found that

Blackhead Persians and Dorper may walk an average of 7,52 km per day compared to 5,31 km by Merino and Karakul.

The average grazing period by sheep is eight hours which corresponds with that of cattle. There are usually 7 grazing periods. The ruminating period varies more than the grazing period and is usually 3/4 hours divided into 9 periods. Most rumination is at night.

Fig. 6.3 Boer Goat

6.4 GOATS

One of the outstanding assets of goats is found in its grazing habits. It is due to these habits that the goat deserves a place in bush, shrub and woodland agriculture. It also makes the goat economically important in these areas. Goats are browsers and their bi-pedal stance gives them an advantage over the sheep with bush and trees. They generally are more active and selective, walk longer distances in search of food and relish a variety of feeds. It will also appear that goats are much less selective in their grazing habits and that certain pioneer grasses and unpalatable shrubs, which sheep refuse to eat, can effectively be used by goats.

Goats can also be used to utilize rough mountainous areas which sheep and cattle avoid. Boer goats and Angora goats graze shrubs to a far greater extent than sheep or cattle. In an experiment on *Acasia karoo,* sheep spent 6 to 10 % of their grazing time on shrub grazing and 50 % on the grass component while Boer goats spent 50 % on the shrub and 50 % on the grass component of the vegetation.

The goat is strongly inclined to graze the vegetation from the top downwards while sheep actually follow the opposite pattern. The goat's utilization level is vertical over a much higher stratum than that of the sheep. This stratum stretches principally from 10 cm to approximately 157 cm.

With the same stocking and with a plentiful supply of bush and grass available, the grazing pressure of goats is much more evenly spread over the camp. Goats can thus play an important part in the management, utilization and prevention of bush encroachment.

6.5 CALCULATIONS OF FLOCK COMPOSITION

The carrying capacity in hectares per small stock unit as determined for each ecological area, is a concrete measure to:
-relate the grazing demands to the actual carrying capacity of the veld.

-calculating the admissible number of stock in terms of the carrying capacity laid down whereby one third or more may be withdrawn.
-determine the admissible number of small stock units which may by kept after withdrawal and which may by exceeded at any time during the year.

The small stock unit is one of the most important norms on which the calculation of flock compositions is based to determine the number of breeding ewes, replacement ewes and lambs for the various farming systems. As far as the calculation of small units (SSU) and large stock units (LSU) are concerned, different workers and countries use different proportional conversion figures. The condition, breed, productive ability, physiological condition, grazing habits etc. all have an impact on the effect that stock have on the grazing and not just their mass.

TABEL 6.2 The following comparative conversion figures are all used in RSA.

TYPES OF ANIMALS AND BREED	SSU	LSU/TYPE	TYPE/LSU
Merino	1,0	0.17	6,0
Dual purpose Merinos	1.1-1.2	0.18-0.20	5.0-5.5
Dorper	1.2	0.20	5.0
Karakul	1.2	0.20	5.0
Angora	0.8	0.13	7.5
Boer goat	1.2	0.20	5.0
Cattle	6.0	1.0	1.0
Horses, donkey	9.0	1.5	1.5

Example :1
Size of farm: 5000 ha, carrying capacity: 5 ha / SSU

Flock statistics:
-Ave. lamb % = 120 %
-Ave. pregnancy rate of ewes = 80 %
-One lambing season per year
-Marking - all lambs except replacement ewes are marketed at age of 4 months
-Replacement = 20 % p.a.
-Young ewes are included in the flock at the age of 12 month
-Ram replacement = 3 %
-Adult rams are purchased and replacement ram lambs are thus not necessary

-Number of Dorper/LSU
-Ram = 4.4
-Dry ewes = 6.5
-Pregnant = 6.0 (5 x 30 days)
-Lactating ewes = 4.5 (4 x 30 day)
-Lamb up to age 4 months = 8.8
-Lambs up to age 12 month = 7.5

Questions:
How many SSU be carried by the farm?
How many SSU grazing days are available on the farm?
The breeding flock consists of 6 age groups with an annual loss of 3%

Determine:
- At what age are the old ewes culled (i.e. no longer used in the ewe flock first young ewes are includes in the flock at age 12 month)
- What percentage replacement ewes are included in the breeding flock at age 12 month
- What percentage lambs must be kept for replacement

- What percentage-breeding ewes can be kept

Calculate the number of ewes that can be kept on 5000 ha to test your answer.

Calculation:

1) $\frac{\text{size of farm}}{\text{carrying capacity}}$ = number of SSU

   - $\frac{\text{5000 ha}}{\text{5 ha /SSU}}$ = 1000 SSU

2) SSU x 365 = 365 000 SSU- grazing days

3) a - 7 years
   b - 18 %
   c - 18 % x 1.03 = 18.5 %

4) determine the grazing days required for a unit of 100 breeding ewes.

3 % rams $\frac{\text{3 x 6.0 x 365 days}}{\text{4.4}}$ = 1493.2 SSU grazing days/annum

5) SSU = $\frac{\text{1 LSU and 4.4 rams – LSU, i.e. a ram is 6.0}}{\text{4.4}}$ = 1.36 SSU

100 ewes – 20 % ewes dry throughout the year (pregnancy rate – 80%)

$\frac{\text{20 x 6.0 x 365 days}}{\text{6.5}}$ = 6738.5 SSU grazing days

80 x of ewes pregnant for 5 month

$\frac{\text{80 x 6.0 x (5 x 30) days}}{\text{6.0}}$ = 12 000 SSU grazing days

80 % of ewes lactating for 4 month

$$\frac{80 \times 6.0 \times (4 \times 30) \text{ days}}{4.5} = 12\,800 \text{ SSU}$$

80 % of ewes dry for 3 month

$$\frac{80 \times 6.0 \times (4 \times 30) \text{ days}}{6.5} = 6\,646.2 \text{ SSU grazing days}$$

120 lambs
100 lamb marketed at 4 month

$$\frac{100 \times 6.0 \times (4 \times 30)}{8.8} = 8\,181.8$$

20 lambs up to 12 month (replacement)

$$\frac{20 \times 6.0 \times 365}{7.5} = 5\,840$$

Total number of SSU grazing days/annum for unit of 100 ewes = 53 699.7
Percentage of the SSU grazing days utilized by the breeding ewes
38 184.7 x 100 = 71.11%
53 699.7

6) 71.11 % of the total grazing days of the farm
= .7111 x 365 000 = 259 551.5 are required by the breeding ewes

7) The total number of breeding ewes which can be maintained by the farm

(100 breeding ewes require 38 184.7 days/annum.

$$\frac{259\,551.5 \text{ grazing days} \times 100 \text{ ewes}}{38\,184.7}$$

= 680 ewes

Test:

3 % rams	- $\frac{.03 \times 380 \times 6 \times 365}{4.4}$	= 10 153.6
Breeding ewe / dry ewes		
Drought ewes	- $\frac{.20 \times 680 \times 6 \times 365}{6.5}$	= 45 821.5
Pregnant	- $\frac{.80 \times 680 \times 6 \times (5x30)}{6}$	= 81 600.0
Lactating	- $\frac{.80 \times 680 \times 6 \times (4x30)}{4.5}$	= 87 040.0
Dry ewes	- $\frac{.80 \times 680 \times 6 \times (3x30)}{6.5}$	= 45 193.8
Lambs marketed	- $\frac{1.00 \times 680 \times 6 \times (4x30)}{8.8}$	= 55 636.4
Replacement	- $\frac{.20 \times 680 \times 6 \times 365}{7.5}$	= 39 712.0
Total SSU grazing days/annum		= 36 5157.3

Example :2
Size of farm: 10 000 ha, carrying capacity: 20 kg/ha

Flock statistics:
-Ave. lamb % = 120 %
-Ave. pregnancy rate of ewes = 75 %
-One lambing season per year
-Marking – all lambs except replacement ewes are marketed at age of 4 months
-Replacement = 20 % p.a.
-Young ewes are included in the flock at the age of 12 month
-Ram replacement = 3 %
-Adult rams are purchased and replacement ram lambs are thus not necessary

Average live mass of Dorper
-Ram ± 82 kg
-Dry ewes ± 52 kg
-Pregnant ± 62 kg (5 x 30 days)
-Lactating ewes ± 48 kg (4 x 30 day)
-Lamb up to age 4 months ± 17 kg
-Lambs up to age 12-month ± 30 kg

Questions:
-How many kg live weight can be carried by the farm?
-What percentage breeding ewes can be kept?
-Calculate the number of ewes that can be kept on the 10 000 ha and test your answer

Calculation:

1) size of farm x carrying capacity = kg/year

- 10 000 ha x 20 kg/ha = 200 000 kg/year

2) Determine the kg/year required for a unit of 100 breeding ewes.

$\frac{\text{3 \% rams } 3 \times 83 \text{ kg} \times 12 \text{ month}}{12} = 246$ kg/year

100 ewes 25 % ewes dry throughout the year (pregnancy rate 75 %)

Dry ♀ $\frac{25 \times 52 \text{ kg} \times 12 \text{ month}}{12} = 1300$ kg/year

Preg. ♀ 75 % of ewes pregnant for 5 months

$\frac{75 \times 62 \text{ kg} \times 5 \text{ month}}{12} = 1937.5$ kg/year

Lac. ♀ 75 % of ewes lactating for 4 months

$\frac{75 \times 48 \text{ kg} \times 4 \text{ month}}{12} = 1200.0$ kg/year

Dry ♀ 75 % of ewes dry for 3 months

$\frac{75 \times 52 \text{ kg} \times 3 \text{ month}}{12} = 975.0$ kg/year

Total for ♀ = 5412.5 kg

120 lambs born

Market: 100 lambs marketed at 4 month

$\frac{100 \times 17 \text{ kg} \times 4 \text{ month}}{12} = 566.7$ kg/year

Replacement 20 lambs up to 12 month (replacement)

$\frac{20 \times 30 \text{ kg} \times 12 \text{ month}}{12} = 600$ kg/year

Total kg/year for a unit of 100 ewes = 6825.2

Percentage "kg" utilized by the breeding ewes

$\frac{5412.5}{6825.2} \times 100 = 79.3$ %

3) 79 % of the total "kg/year" of the farm is breeding ewes = .793 x 200 00

= 158 600 kg/year are required by the breeding ewes on the farm

The total number of breeding ewes which can be maintained by the farm

(100 breeding ewes require 5412.5 kg/year)

$$= \frac{158\ 600.0 \text{ kg/year} \times 100 \text{ ewes}}{5412.5}$$

= 2930 ewes

Test:

3 % rams	- $\frac{.03 \times 2930 \times 82 \times 12}{12}$	= 7 207.8
Breeding ewe / dry ewes		
Dry ewes	- $\frac{.25 \times 2930 \times 52 \times 12}{12}$	= 38 090.0
Pregnant	- $\frac{.75 \times 2030 \times 62 \times 5}{12}$	= 56 768.8
Lactating	- $\frac{.75 \times 2930 \times 48 \times 4}{12}$	= 35 160.0
Dry ewes	- $\frac{.75 \times 2930 \times 52 \times 3}{12}$	= 28567.5
Lambs marketed	- $\frac{1.00 \times 2930 \times 17 \times 4}{12}$	= 16 603.3
Replacement	- $\frac{.20 \times 2030 \times 30 \times 12}{12}$	= 17 580.0
Total kg/year		= 199.977.4
% Stocking		= 99.9

6.6 FACTORS WHICH INFLUENCE THE GRAZING HABITS OF ANIMALS

6.6.1 PERIODICITY

Cattle and sheep graze mainly by day. During long days (summer) they graze in the daytime, provided the temperature is not too high. As the days become shorter more time will be devoted to grazing at night. A close correlation exist between the length of day and the duration of grazing at daylight. The main grazing period begins at daybreak and end as it begins to get hot in mid-day.

Then after drinking water, there is a ruminating session. The second grazing period begins in the afternoon and continues until it is quite dark. As long as conditions in a camp remain constant, the animals follow a remarkably stable routine of activity.

6.6.2 RAINFALL

Rainfall influences the grazing habits of animals both directly and indirectly. Indirectly rainfall influences the grazing habits of animals because it is a determining factor in the nature and conditions of the vegetation. A low rainfall has the effect of producing a sparser cover with more xerophytes hardy types. The grazing periods should thus be longer for the animal to ingest the same amount of food compared to vegetation with a dense cover.

Much time is spent by the animals looking for food. Rain affects the grazing habits of animals directly in that animals cease grazing when it begins to rain and seek shelter. After the rain the animals resume grazing and make up for lost time by eating faster.

6.6.3 TEMPERATURE

High temperature stimulate animals to begin grazing early in the morning and less during the day. It has been established that sheep graze longer on cool days than on hot days. On hot days they graze only early the morning and late afternoon. High temperatures also affect the quality of grazing. On hot days the plants wilt, as a result of poor atmospheric condition. In case of some species, prussic acid poisoning can be caused during wilting.

6.6.4 WIND

Wind mainly affects the evaporation of sweat from the surface of the skin and assists in the regulation of body temperature. On hot days evaporation will take place faster if the wind is blowing and the animal will thus rid itself of excess heat more easily. High/low temperature so often coupled with hot/cold wind, cause a decrease in the total feed intake because the animal feels uncomfortable and seek shelter.

6.7 CROSS-BREEDING

Definition: Is the process during which individuals of two or more breeds are crossed with one another, it allows the breeder to exploit the outstanding traits of two or more breeds to increase production. The resulting offspring is hybrid. Cross-breeding usually results in improved traits in the offspring. Dominant genes tend to mask undesirable recessive genes. The use of cross-breeding generally increases profits from sheep flock.

Cross-bred ewes are hardier, healthier, and produce more milk compared to non cross-bred ewes. Cross-breeding is recommended when producing market lambs. Cross-bred lambs have several advantages over straight-bred lambs.
-they gain weight more rapidly
-are more hardy and vigorous
-have a lower mortality rate

The advantages of using cross-bred ewes instead of straight-bred ewes include:
-greater fertility
-higher lamb survival rate
-higher lambing percentage
-better milk production
-greater ease of lambing
-better maternal instinct
-greater longevity
-better wool quality and higher quantity of wool produced
-early sexual maturity
-greater potential for accelerated lambing
-udder soundness

Fig. 6.4 Cross-breed (Dor x Kar)

Characteristics to look for when selecting the breed of ram to use in cross-breeding program, include:
-rapid growth
-good carcass quality
-greater sexual aggressiveness
-above-average testicle size at puberty
-high fertility
-high survival rate of lamb offspring

6.7.1 CROSS-BREEDING SYSTEMS

Genetic considerations, to which attention should be paid when compiling a cross-breeding system is:

-The magnitude of the difference between breeds.

Crosses between breeds can be used for:

-Upgrading of one breed.
-Development of a system where one breed complements the other in respect of economically important production.
-Development of a new breed.

CHAPTER 7: FACILITIES FOR SMALL STOCK

7.1 INTRODUCTION

All improvements on a farm should only be implemented after good planning. Temporary or ineffective improvements cost more money in the long run. When improvement is made, attempts should be made to provide the animal with as much comfort as possible, and in the second place, deteriorating labour situation should be borne in mind. One should also keep future expansion in mind.

7.2 MAKING CAMPS

Fig. 7.1 Stock proof fence

The type of camps and environmental conditions will determine the nature and number of camps. Stud breeders need more small camps than flock farmers. The availability of water, the nature of the veld, the presence of predators, the method of tending and capital normally determine the size of the camps. In the mountainous areas the camps must be smaller so as to make the rounding up of the animals easier.

In sandy regions and poor veld the camps can be larger. In all cases a few small camps near the farmstead are very handy. Before camps are made, the farmer must make sure that sufficient water will be available. The animals should not have to walk too far to reach the watering places.

Fig. 7.2 Jackal proof fence

7.3 WATERING FACILITIES

Clean water must be available at all times, this is only possible where water is provided in proper drinking troughs. Open water and puddles of water next to drinking troughs is a source of parasites and foot rot infection. The troughs must be cleaned regularly with hard scrubbing brushes. Troughs must not be too small, otherwise the animals will empty it quickly and then start trampling each other. An effective trough has a drinking space of 15-20 cm and is approximately 22.5 cm deep, the length depends on the number of animals. Each animal needs 30 cm of space. Small stock drinks about 2.5 to 4.5 liters of water per day. The paving of the area around the trough is recommended so as to provide a good foothold for the animals. The trough must be raised above the ground level to prevent contamination with manure.

Fig. 7.3 Water trough

7.4 CONSTRUCTION OF KRAALS

Effective kraals at central outposts eliminate the driving of animals over long distances. Kraals in the faraway areas of the farm must be constructed in such a way that they can serve three or four camps simultaneously. The best place is close to a water point. It is advisable to build kraals on high ground to facilitate drainage, particularly in regions with high rainfall. Before a kraal is designed, the number of small stock must be carefully considered, chasing animals around in unnecessary large kraals in order to catch them, is highly undesirable, whereas too small kraals are equally undesirable.

The following may serve as a guideline:

-0.28 m^2 per animal: then the animals are closely packed.
-0.37 m^2 per animal: then the animals stand comfortably.
-0.46 m^2 per animal: then they have room to move around.

Gates at kraals must be neat. The common practice is to always place the gate along the length of the fence, so that it can open flat against the fence, and it must always open in the direction of the moving animals.

The gates must also be wide enough to allow a trailer to pass through for removal of manure, to help prevent ectoparasites.

The kraals must be suitable for sleeping and for routine treatment such as vaccinations, dosing and sorting. These pens become more and more important on account of the increasing labour shortage. Less manpower is required and fewer injuries occur, particularly during dosing, vaccination, sorting or when catching animals.

The dosing pen should not be too long, because the animals in the rear can be crushed easily. A long dosing pen can however be made if it is provided with partitions. The normal dimensions of a dosing pen are 9-10 m long, 0.9 – 1.2m wide and approximately 1 m high.

The sorting pen is very handy when taking out pregnant ewes or when animals need to be sorted for some or other reason. For this purpose the length should not be less than 9 m. The width should be 40 – 45 cm, 1.2 m high, so as to allow only one animal to pass, and not turn around.

Fig. 7.4 The kraals must be suitable for sleeping and for routine treatment such as vaccinations, dosing and sorting.

7.5 DIPS

Routine dipping for the control of ectoparasites should be included in every management program. Various methods of dipping exist and for submersion of the animals there are round and oblong dipping tanks. The round tank is generally used and the advantage of this tank is that it can be erected cheaply and also takes less water. The disadvantage is that the animals have to be caught individually and placed in the water.

Another method of dipping is spray race. The equipment is cheaper and easier to dismantle and transport if necessary. It requires only slightly more water that the tank. If a motor pump is used, however, the apparatus become more expensive than in the case of tanks. From 35 to 50 sheep are herded into the round spray kraal and are then sprayed with dipping fluid from all directions for a certain period. The advantage of this method is that it saves a considerable amount of labour and that ewes in lamb can be dipped safely.

During the rainy season small stock must be herded through a foot dip to control foot rot and ticks on the feet of the animal. A foot dip can be installed at very little extra cost in the kraal system and the animals can be herded through regularly as and when required.

Fig. 7.5 Spray dip

7.6 LICK TROUGHS

A salt phosphate lick must be available to the animals at all times. A lick trough should be provided in every kraal or camp.

Fig. 7.8 Lick trough

7.7 SHED / SHELTER

Although a shed is not absolutely essential, it is very convenient on any farm. Such a shed can accommodate feed troughs, a shearing board, serving pens for hand-serving or I.A., and can also serve as a room for storage of medicine and equipment. Trees should be planted on barren farms to provide shade, especially close to the water points.

Fig. 7.9 Sheep shed and feed troughs

7.8 PROBLEM ANIMAL CONTROL

7.8.1 INTRODUCTION

There are a number of animals which may at times cause damage to livestock.There are two officially-proclaimed problem animals, viz. the black-backed jackal and the caracal (Ordinance 14 of 1978). Domestic dogs and caracal (lynx) are capable of killing fully grown sheep.

7.8.2 IDENTIFYING THE CULPRIT

Unless the culprit is correctly identified from the killing and feeding signs, a lot of effort may be wasted by using inappropriate methods to capture the stock killer.

7.8.2.1 Black-backed jackal

- Kills mainly lambs
- Only one killed at a time
- Throat bite shows two small punctures on either side of windpipe.
- Average distance between upper canines 25 mm (range: 23-29 mm).

-When feeding, the jackal opens the carcass on flank between hip and bottom of ribs.
-Eats a small amount. Parts eaten are usually kidneys, liver, heart, and tips of ribs.
-Carcass not moved from spot where killed.

Fig. 7. 10 Black-backed jackal

7.8.2.2 Caracal (lynx)

-Kills sheep or goat
-One killed at a time
-Either a throat bite with two small punctures on either side of windpipe or a bite at the back of neck also shows two punctures on either side of spine.
-Average distance between upper canines 25 mm (range: 24 -30 mm).
-Claw marks on shoulder visible sometimes.
When feeding, the caracal feeds on flesh between the hind legs or on the inside of a hind leg. Sometimes the brisket is eaten.

- If it returns to the carcass, eats mainly soft flesh such as the shoulder, sometimes tips of ribs but never large bones.
- Prey sometimes partly covered by scratching soil or plant debris over it after feeding.

Fig. 7. 11 Caracal (lynx)

7.8.2.3 Dogs

-Either lambs or adult sheep are killed. Usually more than one sheep is killed or maimed, but occasionally only one.

-When very small lambs are killed, they are bitten across the chest, back, or head.

-When sheep or large lambs are killed, they are bitten anywhere on the body, e.g. hind legs (rump or hamstrings), stomach or flank ripped open. Sometimes the neck, or ears are bitten.

-Kill not as neat and clean as that of either jackal or lynx.

-When feeding, dogs sometimes eat little or even nothing, but on other occasions large amounts are eaten (more than that eaten by jackal or lynx).

-Usually start feeding from rear, ripping off large pieces and also eating large bones.

7.8.3 CONTROL MEASURES FOR CARNIVORES

All control measures are not equally effective for all predators. The effectiveness of control methods has been evaluated for selectivity, cost and availability, time and effort involved in its use, specialist knowledge needed to use it, and how effective it is in killing or capturing the culprit.

The most suitable methods of control for the five predators on small livestock are:

Black-backed jackal - gin traps
Caracal - cage traps, gin traps
Dog - cage trap, gin traps

7.8.4 GIN TRAP / SLAGYSTER

Gin traps, also known as jawtraps, leg-hold traps, or slagysters, have proved that they can be effective for problem-animal control, provided they are carefully set and sited. The following guidelines should be followed with care: Traps should be set beside paths and tracks which are being used by the predators. Never set a trap in a path as this increases the chances of catching non-target animals. Do not bury the trap too deeply. A covering of 3 mm over the pan and jaws is sufficient. If buried too deeply the jaws will not close rapidly. Set the trap beside the path which the lynx has been using. If a carcass is found which the lynx has partly buried or dragged under a bush, the trap may be set beside it, as the lynx may return to the kill.

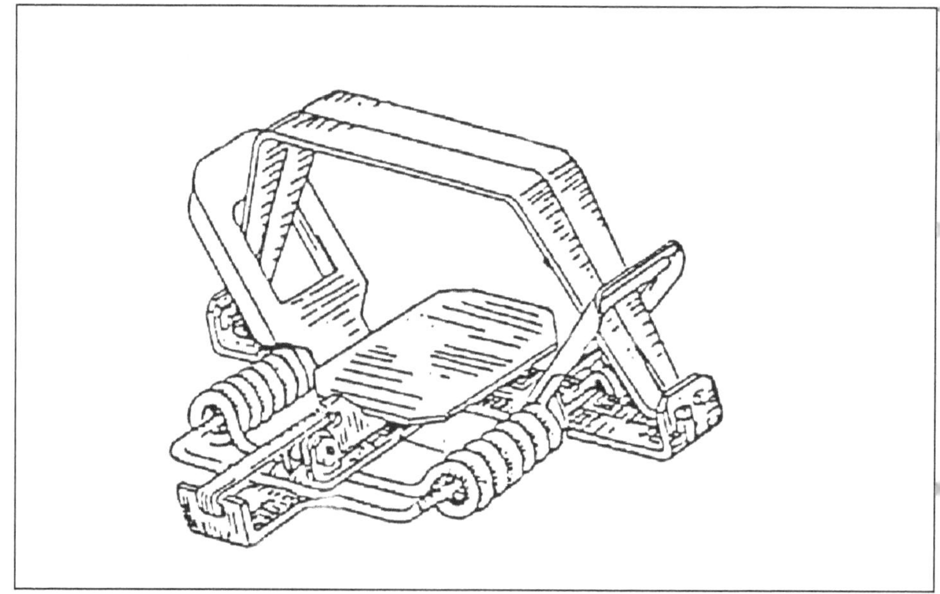

Fig. 7.12 GIN TRAP / SLAGYSTER

7.8.5 CAGE TRAP FOR LYNX

The back of the trap should be pushed against a bush and branches should be laced on top of the trap and along the sides: Cover the floor and trigger plate with soil to give a natural appearance. Place about three spoonfuls of bait in a tin which should be closed, with holes punched in the lid or around the top. The tin should be hung at the back of the trap.

Bait should be made from 1 cup fishmeal, 1 cup blood, and 1 cup mince, which should be thoroughly mixed. Store the bait in a tightly sealed bottle,which should be buried if flies become a nuisance. The bait is ready for use about one week after mixing it.

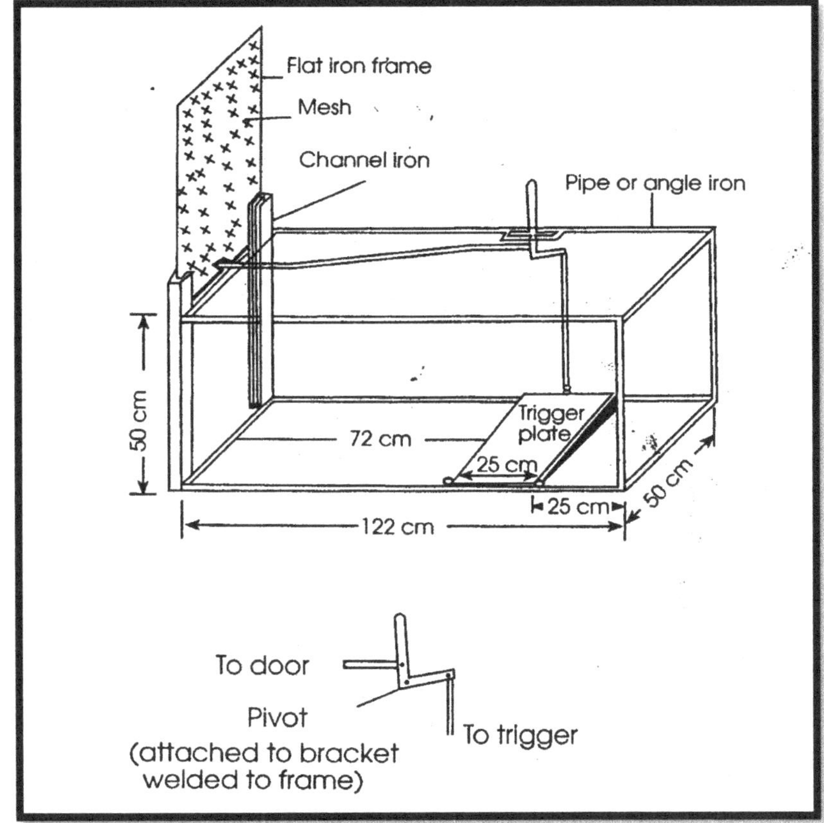

Fig. 7. 13 CAGE TRAP FOR LYNX

CHAPTER 8: RECORD KEEPING

8.1 WHAT IS MEANT BY RECORD KEEPING?

Stock record keeping is the collecting and recording of data and information about individual animals or groups of animals, the analysis of the collected data, and the use of the analysed data and information for profitable and more effective stock farming. Stock records form the foundation of modern stock farming. With very high input costs the farmer cannot afford to carry unproductive animals on his farm.

Records will enable him to identify and select individual animals that adapt well under his management, and to cull poor-performing animals.

Management mistakes and wrong decisions can be identified. Financial records are more accurate and meaningful if they are related to animal performance. Records are taken in the veld and transferred in the office to record books or to the computer. For veld recording, the back of a cigarette or matchbox will not do.

8.2 ANIMAL IDENTIFICATION

It is essential that animals can be identified. If animals can be individually identified then record systems can be improved to suit specific needs, and can be expanded by adding on. Examples of ways in which animals can be identified are as follows:

Tags can be used to identify individual animals within groups of animals.

Tag single lambs on left ear, tag facing forward or up.
Tag twin lambs on right ear, tag facing forward or up.
Tag triplets on right ear, tag facing back or down.
Tag quadruplets on left ear, tag facing back or down.

Stencil the ewe number on the back of the metal tag which is used to tag the ewe's lamb. A lost lamb's mother can then be traced quickly. Use coloured leg tags to identify ewe and lamb groups, e.g. twins and singles,ewe and ram mating groups and their offspring.

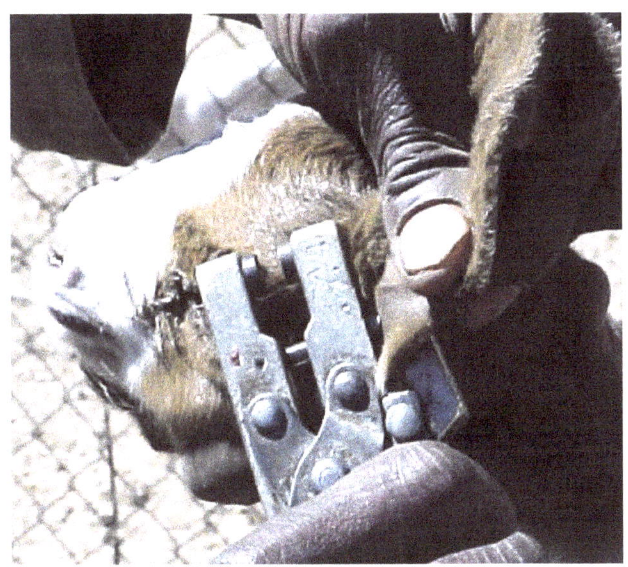

Fig. 8.1 Mark of lambs

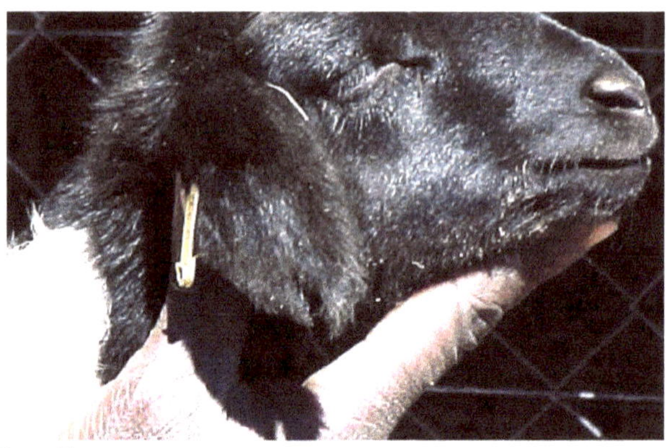

Fig. 8.2 Metal eartag

8.3 IDENTIFICATION AND ON - SHEEP / GOAT RECORDS

Very basic records can be kept on the sheep. Here no bookwork is involved. Management must be of the highest order, or this system could be used to the detriment of the flock. The on-sheep record system might appeal to the more extensive sheep farmer who keeps no records at all. The ears of the sheep are the record book.

8.4 RECORD BOOKS

8.4.1 FIELD BOOKS

8.4.1.1 THE STOCK BOOK

- This book forms the backbone of the record-keeping system and is kept by the shepherd or the farmer,
- the following information is recorded daily:
    - -Livestock counts
    - -Livestock movements
    - -Livestock deaths
    - -Number of sick animals and treatments

FOR EXAMPLE:

-Camp W 10, 210 ewes, 2 rams, and 2 ewes death = vermin
-Camp W 11, 70 cows
-Camp W 16, 30 weaned ewe lambs, dose (12/02/03) – Agvetgreen

8.4.1.2 MATING BOOK

This is used to record all mating in the veld. This book enables the farmer to group ewes lambing down in the same week. Attention can then be focused on the supervision of the animals lambing down during a certain period.

TABLE 8. 1 MATING BOOK - NOVEMBER 2000

DATE	EWE NO	RAM NO	EWE MASS - KG	EWE C.S *	REMARKS
11/01/00	8822	CL 323	65	4.0	
11/01/00	8878	CL 300	56	4.0	
11/01/00	8830	CL300	59	2.5	Teeth, ewe thin
12/01/00	8835	CL 323	80	5.0	Over fat

* C.S.-Condition score

8.4.1.3 Lambing book

This book is to record all lambing in the veld.

TABLE 8. 2 LAMBING BOOK APRIL – 2002

DATE	EWE NO	EWE MASS. KG	EWE C.S.	LAMB NO - KG	REMARKS
01/04/02	5434	70.0	4.5	702 (r)	TW
	5434		4.0	703 (e)	
02/04/02	5430	60.0	4.0	703 (r)	S
02/04/02	5469	60.0	5.0	704 (e)	S

* C.S. - Condition score

8.5. OFFICE RECORDS

8.5.1 DEATH REGISTER

Deaths are copied from the stock book to the register every day and ended off at the end of each month. With these records one can pinpoint the times of the year when more management input is required to minimize stock losses.

TABLE 8. 3 DEATH REGISTER

DATE	ANIMAL NO.	P.M.	REMARKS
20/04/02	4490	No	Poison plant
02/05/02	4302	Yes	Pasteurellosis
26/12/02	4190	No	Poor condition

8.5.1.1 ANIMAL REGISTER / STOCK LIST

List all female tag numbers, tag in numerical order, on an age basis, starting with the oldest. Date and cause of death must be recorded.

8.5.1.2 BREEDING RECORD BOOK FOR RAM / EWE

This book contains information about the fertility of the ram or ewe and the breeding performance.

TABEL 8. 4 DAMARA SHEEP EWES LAMB, NOV. - 2002

Birth	Lamb	EWE	RAM	Birth	Ram	Ewe	S	TW	T	Q	Pre\W-	Still-	Fo-	Fos-	Des-	EWE	Lamb
Day	No.	No.	No.	weight							MO	BORN	LAMB	EWE	LAMB		
13-Nov-02	1521	625	122	5.0		Ewe	S									625	1521
13-Nov-02	1222	1260	122	4.5	Ram		S									685	1527
14-Nov-02	1523	951	122	4.5	Ram		S									951	1523
14-Nov-02	1524			4.5		Ewe	S						*			1260	1222
14-Nov-02	1525			3.0	Ram		S						*				1524
14-Nov-02	1526			3.0		Ewe	S						*				1525
15-Nov-02	1527	685	120	3.0	Ram		S										1526
	7			3.9	4	3	7	0	0	0	0	0	3	0	0		

8.5.1. GENERAL

What is needed to achieve one's goals?

- Simple and permanent identification system.
- Records of birth mass.
- Records of milk production of ewe.
- Records of growth potential of lambs.
- Records post weaning growth of the lambs.
- Records of ram performance.

BIBLIOGRAPHY

ANIMAL PRODUCTION IV. Study Guide 5, 6, 1ANP411AE., Technikon S.A. Florida. 1/2000.

ALLDEN, W.G., 1979. *Feed intake, diet composition and wool growth.*

BOSHOFF, D.A., 1980. *Die invloed van laktasie en seisoene op geslagsaktiwiteit en die kunsmatige sinchronisasie van oestrus by lakterende Karakoelooie.* S.Afr. Tydskr. Veek.5,37.

CHURCH, D.C., *Livestock Feeds & Feeding,* Third Edition.

CLAY, P., MATHIS, TIM ROSS., 2000. *Using Hormones to Control Reproduction in Sheep.*

COLAS, G., 1980. Variations saisonnière de la qualité du sperme Chez le belier Ile-de-France. I. Etude de la morphologie celluliare et de la motilité massale.
Reproductio, Nutrition, Développement 20, p 1789-1799.

COLAS, G., 1981. Variations saisonnière de la qualité du sperme chez le belier Ile-de-France. II Fécondance: relation avec les critères qualitifs observes in vitro.
Reproductio, Nutrition, Développement 21, p 399-407.

DE WET, J., EN GARETH BATH, *Small stock Diseases.,* 1994.

GILLESPIE, JAMES R. *Modern Livestock & Poultry Production*, Sixth Edition.

GRANDIN, T., *Livestock handling and transport,* CABI Publishing.

GREYLING, J.P.C., Greeff, J. C., Brink, W.C.J., Wyma, G.A., 1988. *Synchronization of estrus in sheep of low-narmal mass under range conditions. The use of different progestagen and PMSG.*, South African Journal of Animal Science. Volume 18, (4) p 164-176.

GREYLING, J.P.C., Kotze, G.J.,Taylor G.J., Hagendijk W.J., 1993. *Synchronization of estrus in sheep outside the breeding season.*, South African Journal of Animal Science. Volume 24, (1) p 33-37.

HERBST, I.A., VAN DER WESTHUIZEN, J.M., SCHUTT, A.P., & STEYN, J.J. *Reproduksie en K.I. by kleinvee*. 1990., Deel 1-10.

HAYNES, N.B., SCHANBACHER, B.D. 1983. *The control of reproductive activity in a ram.*

KURSES VIR LANDBOUKOLLEGESTUDENTE 1983., *Dorperskape en Karakoelskape.*, Departement van Landbou,. Neudamm, 8: 1-10., 9: 1-6.,10: 1-9.,17:1-24., 18:1-6.,20: 6.

LANDBOU WEEKBLAD 1999., *Entingsprogramme vir Skape* (5)., 1-2.

LANDBOU WEEKBLAD 1999.,*Verseker 'n hoër besettingsyfer.*,1-5

KORVER, S., & ARENDONK VAN J.A.M. EDITORS., *Modelling of Livestock Production Systems.*

KURSES VIR LANDBOUKOLLEGESTUDENTE 1983., *Grondbeginsels van veld- en weidingbestuur.*, Departement van Landbou Weidingsleer I, 7:116., 9:142-152.

KURSES VIR LANDBOUKOLLEGESTUDENTE 1984., *Beginsels van veldbestuur.*, Departement van Landbou Weidingsleer II, 1: 1, 4: 72-74, 7: 117.

KURSES VIR LANDBOUKOLLEGESTUDENTE 1983., *Grondbeginsels van veld- en weidingbestuur.*, Departement van Landbou Weidingsleer III, 1: 5.

MAREE, C., & CASEY, N.H., 1993. *Livestock Production Systems Agri.- Development Foundation.,* 7: 124-148.

MCDONALD, EDWARDS, GREENHALGH, MORGAN, *Animal Nutrition,*
Fifth Edition.

MőNNIG, H.O., EN VELDMAN, F.J., *Handboek oor Veesiektes.,* 1996.

NEUDAMM AGRIC. COLLEGE, *Small stock Production.,* 2000.

NEUDAMM AGRIC. COLLEGE, *Diseases of Sheep and Goats,* 2000.

PEARSE, D., OLDHAM, C.M., 1994. *Reproduction in sheep.* Cambridge University Press, Cambridge, p 24-26.

ROBINSON, T.J., 1988. *Controlled sheep breeds*: update 1980-1985.Aust. J. Biol. Sci. 41, p 1-13.

SA STUD BOOK ASSOCIATION – Bloemfontein. July 2006.

SACHSE JAMES, M., 1996, *Sheep Production and Management.*, College of Agriculture and Home Economics New Mexico State University

SALAMON, S., and ROBINSON, T.J., 1962. Studies on AI of Merino sheep:
The effect of frequency and season of AI, age of the ewe, ram and milk diluents on lambing performance. *Australian Journal of Agric. Research* 13, p 52-68.

SKINNER, D.J., and ROWSON, L.E.A., 1968. Puberty in Suffolk and cross-bred rams. *Journal of Reproduction and Fertility* 16, 479-488.

STELLFLUG, J.N., 2002. *Influence of classification levels of ram's sexual activity on Spring breeding ewes.* USDA-ARS US Sheep Experiment Station,
USA.

SUNVET., 1995. *Animal Health.*

THERON, E.P., HARWIN, G.O., 1976. *The role of intensive pastures for livestock production.* RSA Journal of Animal Science 6(2): 139.

TURNBULL, D., 1997. *Journal of Reproduction and Fertility.* 7. p 207.

QUIRKE, J.F., 1979a. Estrus ovulation, fertilization and early embryo mortality in progestagen-PMSG treated Galway ewe lambs. *Irish Journal of Agricultural Research* 18, p 1-11.

QUIRKE, J.F., 1979b. Control of reproduction in adult ewes and estimation of reproductive wastage in ewe lambs following treatment with progestagen impregnated sponges and PMSG. *Livestock Production Science* 6, p 295-305.

ROBERT, E., TAYLOR, THOMAS, G., FIELD, *Scientific Farm Animals Production,* An Introduction to Animal Science, Sixth Edition.

VOSTER, L., F. *Weidingkunde Deel A, WDL 338.* 4: 97-107 , 11: 140-142, 13: 196-198.

WENNBOM, D., 1994. *Journal of Reproduction and Fertility.* 7. p 205.

INDEX

N

O

P

R

S

Management Livestock Calendar- Region Based

Area : **Eastern Cape**
Type of Grazing : **Veld**
Type of Livestock : **Angora goats**
Author : **Jansenville**
First Mating month : **4** Second Mating month:
First Marketing age: **18** Second Marketing age:

Month	Activity
Jan.	Dose all goats broad spectrum anthelmintic
	Kids dose broad spectrum incl. tape worm (4 month-old lambs)
	Kids Pulpy kidney 2 + Pasteurella 2 (4 month-old lambs)
	Wean kids (4 month-old lambs)
Feb.	Ewes, rams and young goats Pulpy Kidney + Pasteurella
	Ewes Ensootic Abortion (April/May mating group)
	Finalise mating ewe flock (April/May mating group)
	Flushing mating flock if required (April/May mating group)
	Ram kids to be kept as sires - Rev1 (5 month-old lambs)
Mrch.	Fertility testing of rams (April/May mating group)
	Kids Rift Valley fever + Wesselsbron disease (6 month-old lambs)
	Market surplus animals (18 month-old lambs)
Apr.	6 to 8 week Mating period (April/May mating group)
May	
Jun.	Dose all goats broad spectrum incl. nasal worm
	Scan ewes for pregnancy 42 days after mating (April/May mating group)
Jul.	Supplement ewes last 6 to 4 weeks of pregnancy if required (September/October lambing group)
Aug.	All goats Pulpy kidney + Pasteurella
	Supplement ewes last 6 to 4 weeks of pregnancy if required (September/October lambing group)
Sept.	Ewes are kidding (September/October lambing group)
	Supplement lactating ewes if required (September/October lambing group)
Oct.	Dose all goats broad spectrum
	Supplement lactating ewes if required (September/October lambing group)
Nov.	Kids 8 weeks : Pulpy Kidney 1 + Pasteurella 1 (2 month-old lambs)
	Kids 8 weeks : Dose against Milk tape worm (2 month-old lambs)
Dec.	

Comments
Apart from the strategic drenchings mentioned in the programme, feacal egg counts should be done regularly to determine if additional drenchings are necessary, especially in the rainy season.

Area : **Eastern Cape**
Type of Grazing : **Veld**
Type of Livestock : **Sheep**
First Mating month : **4** Second Mating month:
First Marketing age: **10** Second Marketing age:

Month	Activity
Jan.	Wean lambs (4 month-old lambs)
	Dose lambs broad spectrum (4 month-old lambs)
	Inoculate ewe lambs against Enzootic abortion (4 month-old lambs)
	Inoculate all ewes and rams Pulpy kidney and Pasteurella (4 month-old lambs)
	Supplementary feeding for lambs - 2 weeks after weaning (4 month-old lambs)
Feb.	Select rams on performance testing and do fertility tests (April/May mating group)
	Dose rams and ewes six weeks before mating (April/May mating group)
	Control external parasites (April/May mating group)
	Supplementary feeding for 2-tooth ewes and rams if necessary (April/May mating group)
	Cut horns and trim hooves of rams (April/May mating group)
	Class ewes on production, reproduction and age (April/May mating group)
March	Crutch all ewes (April/May mating group)
	Put 3% teaser rams between ewes - 14 days before mating (April/May mating group)
	Exercise breeding rams (April/May mating group)
	Flush feeding, supplementary feeding or licks for ewes (April/May mating group)
	Control external parasites (April/May mating group)
April	Remove teaser rams (April/May mating group)
	Use 3% acitve and fertile rams (April/May mating group)
	Mate two-tooth ewes to 4% experienced rams (April/May mating group)
	6 Week mating period (April/May mating group)
	Continue flush feeding (April/May mating group)
	External parasite control (April/May mating group)
May	Continue with supplementary feeding if necessary (April/May mating group)
June	Scan ewes for pregnancy 42 days after mating (April/May mating group)
	External parasite control (April/May mating group)
	Keep pregnant ewes with twins separate (April/May mating group)
	Identify dry ewes (April/May mating group)
	Cull ewes who have missed for the second time (April/May mating group)
	Continue with supplementary feeding if necessary (April/May mating group)
July	Young ewes Blue udder 1 (September/October lambing group)
	Dose ewes with broad spectrum (September/October lambing group)

Aug.	Supplementary feeding last 4 weeks pregnancy if necessary (September/October lambing group)
	Ewes Blue udder, young ewes Blue udder 2 (September/October lambing group)
Sept.	Ewes are lambing (September/October lambing group)
	Supplementary feeding lactating ewes if necessary (September/October lambing group)
Oct.	Supplementary feeding lactating ewes if necessary (September/October lambing group)
	External parasite control (September/October lambing group)
	Dose for round and tape worms (1 month-old lambs)
	Castrate and dock tails of lambs (1 month-old lambs)
Nov.	
Dec.	

Area : **Eastern Karoo**
Type of Grazing : **Veld**
Type of Livestock : **Merino sheep**
Author : **Grootfontein ADI**
First Mating month : **4** Second Mating month: **10**
First Marketing age: **10** Second Marketing age: **12**

Month	Activity
Jan	Blue tongue C 4 months old lambs third week January
	Young ewes Blue udder 1 (March/April lambing group)
	Wean lambs + dose broad spectrum incl. tape worm (4 month-old lambs)
	Rev1 ram lambs to be kept as sires (4 month-old lambs)
	Market lambs that are ready (10 month-old lambs)
Feb.	All sheep Pulpy kidney
	Suppl feeding last 4 weeks pregnancy if required (March/April lambing group)
	Ewes Blue udder; young ewes Blue udder 2 (March/April lambing group)
	Rift valley fever + Wesselsbron lambs kept for breeding (5 month-old lambs)
	Market lambs that are ready (11 month-old lambs)
March	Mating flock Enzootic abortion (April/May mating group)
	Finalise mating flock (April/May mating group)
	Mating flock flushing if required (April/May mating group)
	Rams fertility testing (April/May mating group)
	Put in teaser rams 14 days prior to mating (April/May mating group)
	Ewes are lambing (March/April lambing group)
	Supplementary feeding lactating ewes if required (March/April lambing group)
	Market lambs that are ready (12 month-old lambs)
April	6 week mating period (April/May mating group)
	Market old ewes (April/May mating group)
	Don't work with mated ewes unnecessarily (April/May mating group)

	Supplementary feeding lactating ewes if required (March/April lambing group)
	Docking and castrating of lambs (1 month-old lambs)
	Market lambs that are ready (13 month-old lambs)
May	Don't work with mated ewes unnecessarily (April/May mating group)
	Lambs 8 weeks : Pulpy kidney 1 + dose tape worm (2 month-old lambs)
	Market lambs that are ready (8 month-old lambs)
	Market remainder of lambs (14 month-old lambs)
Jun.	Blue tongue B rams fourth week in June
	Dose all sheep nasal worm
	Blue tongue A rams fisrt week in June
	Scan ewes for pregnancy (April/May mating group)
	Lambs 12 weeks : Pulpy kidney 2 (3 month-old lambs)
	Market lambs that are ready (9 month-old lambs)
Jul.	Blue tongue A ewes + all lambs first week July
	Blue tongue C rams third week in July
	Blue tongue B ewes + all lambs fourth week July
	Young ewes Blue udder 1 (September/October lambing group)
	Wean lambs + dose broad spectrum incl. tape worm (4 month-old lambs)
	Rev1 ram lambs to be kept as sires (4 month-old lambs)
	Market lambs that are ready (10

	month-old lambs)
Aug.	All sheep Pulpy kidney
	Blue tongue C ewes + all lambs third week August
	Suppl feeding last 4 weeks pregnancy if required (September/October lambing group)
	Ewes Blue udder; young ewes Blue udder 2 (September/October lambing group)
	Rift valley fever + Wesselsbron lambs kept for breeding (5 month-old lambs)
	Market lambs that are ready (11 month-old lambs)
Sept.	Mating flock Enzootic abortion (October/November mating group)
	Finalise mating flock (October/November mating group)
	Mating flock flushing if required (October/November mating group)
	Rams fertility testing (October/November mating group)
	Put in teaser rams 14 days prior to mating (October/November mating group)
	Ewes are lambing (September/October lambing group)
	Supplementary feeding lactating ewes if required (September/October lambing group)
	Market remainder of lambs (12 month-old lambs)
Oct.	6 week mating period (October/November mating group)
	Market old ewes (October/November mating group)
	Don't work with mated ewes unnecessarily (October/November mating group)
	Supplementary feeding lactating

	ewes if required (September/October lambing group)
	Docking and castrating of lambs (1 month-old lambs)
Nov	Don't work with mated ewes unnecessarily (October/November mating group)
	Lambs 8 weeks : Pulpy kidney 1 + dose tape worm (2 month-old lambs)
Dec.	Blue tongue A 3 months old lambs first week December
	Dose all (except weaner lambs) broad spectrum + nasal worm
	Blue tongue B 3 months old lambs fourth week December
	Scan ewes for pregnancy (October/November mating group)
	Lambs 12 weeks : Pulpy kidney 2 (3 month-old lambs)

Comments
NB Bluetonge vaccination of rams: If you only have an autumn mating season, rams could receive their Blue tongue vaccinations together with the ewes and lambs from the first week in Juy. If you also have a spring mating season, rams should get their Blue tongue vaccinations earlier, in order to allow a 8 to 12 week period between the last vaccination and the start of the mating season.

Area : **Karoo : Autumn mating**
Type of Grazing : **Veld**
Type of Livestock : **Sheep**
Author : **Louis Du Pisani, NWGA**
First Mating month : **4** Second Mating month:
First Marketing age: **9** Second Marketing age:

Month	Activity
Jan.	Ewes Blue tongue C
	Lambs 16 weeks old - Blue tongue C, Pulpey kidney 2 (4 month-old lambs)
Feb.	Mating ewe flock Enzootic abortion, Blue udder (April/May mating group)
	Mating ewe flock Broad spectrum anthelmintic + Nasal worm (April/May mating group)
	Mating ewe flock Multimin, Vit A (April/May mating group)
	Keep rams fit by ensuring enough exercise (April/May mating group)
	Rams flushing, Multimin, Vit A (April/May mating group)
	Rams broad spectrum + Nasal worm (April/May mating group)
	Lambs 20 weeks - Wean (5 month-old lambs)
	Lambs 20 weeks - Multimin, Vit A (5 month-old lambs)
March	Dry ewes & Rams Pulpey kidney & Pasteurella
	Mating ewe flock flushing (April/May mating group)
	Crutch long wool ewes before mating (April/May mating group)
	Cull ewes with bad udders/teats and ewes which did not lamb (April/May mating group)
	Teaser rams 14 days before mating (fat tail type rams) (April/May mating group)
	Test rams for fertility, veneric diseases & mating ability (April/May mating group)
	Ewe lambs - Rift Valley Fever, Wesselsbron disease (6 month-old lambs)
April	6 week mating period (April/May mating group)
	Market lambs that are ready (7 month-old lambs)

May	Avoid unnecessary working with mated ewe flock (April/May mating group)
	Market lambs that are ready (8 month-old lambs)
June	Ewes Nasal worm
	Rams Blue togue A, Nasal worm
	Scan ewes for pregnancy (April/May mating group)
	Market lambs that are ready (9 month-old lambs)
July	Rams Blue tongue B, Pasteurella
	Market lambs that are ready (10 month-old lambs)
Aug.	Rams Blue tongue C, Pulpey kidney
	Ewes late pregnancy - Broad spectrum anthelmintic (September/October lambing group)
	Ewes late pregnancy - Multimin, Vit A (September/October lambing group)
	Ewes late pregnancy - Multivax P (September/October lambing group)
	Market remainder of lambs (11 month-old lambs)
Sept.	Rams Multimin, Vit A
	Ewes are lambing (September/October lambing group)
Oct.	
Nov.	Ewes Blue tongue A
	Lambs 8 weeks old - Blue tongue A, Milk tapeworm (2 month-old lambs)
Dec.	Ewes Blue tongue B
	Lambs 12 weeks old - Blue tongue B, Tape worm (3 month-old lambs)
	Lambs 12 weeks old - Pulpey kidney 1, Pasteurella (3 month-old lambs)

Area : **Karoo : Spring mating**
Type of Grazing : **Veld**
Type of Livestock : **Sheep**
Author : **Louis Du Pisani : NWGA**
First Mating month : **11** Second Mating month:
First Marketing age: **8** Second Marketing age:

Month	**Activity**
Jan.	Scan ewes for pregnancy (November/December mating group)
	Market lambs that are ready (9 month-old lambs)
Febr.	Market remainder of lambs (10 month-old lambs)
March	Rams Multivax P
	Ewes late pregnancy - Multimin, Vit A (April/May lambing group)
	Ewes late pregnancy - Multivax P (April/May lambing group)
	Ewes late pregnancy - Broad spectrum + Nasal worm (April/May lambing group)
April	Ewes are lambing (April/May lambing group)
May	
Jun.	Ewes & Rams - nasal worm
	Rams - Blue tongue A first week in June
	Rams - Blue tongue B fourth week in June
	6 weeks old lambs - Milk tapeworm (2 month-old lambs)
Jul.	Ewes - Blue tongue A first week July
	Ewes - Blue tongue B fourth week July
	Rams - Blue tongue C third week in July
	12 weeks old lambs - Tape worm

	(3 month-old lambs)
	12 weeks old lambs - Pulpey kidney 1, Pasteurella (3 month-old lambs)
Aug.	Ewes - Blue tongue C third week August
	16 weeks old lambs - Pylpey kidney 2 (4 month-old lambs)
Sept.	Mating ewe flock - Enzootic abortion, Blue udder (November/December mating group)
	Mating ewe flock - Multimin, Vit A (November/December mating group)
	Mating ewe flock - Broad spectrum + Nasal worm (November/December mating group)
	Keep rams fit by ensuring enough excercise (November/December mating group)
	Rams flushing, Multimin, Vit A (November/December mating group)
	Rams Broad spectrum + Nasal worm (November/December mating group)
	20 weeks old lambs - Wean lambs (5 month-old lambs)
	20 weeks old lambs - Multimin, Vit A, Blue tongue A (5 month-old lambs)
Oct.	Ewes & Rams - Pulpey kidney, Pasteurella
	Mating ewe flock flushing (November/December mating group)
	Crutch long wool ewe before mating (November/December mating group)
	Cull ewes with bad udders/teats and ewes which did not lamb (November/December mating
	group)
	Teaser rams 14 days before mating (fat tail type rams) (November/December mating group)
	Test rams for fertility, veneric diseases & mating ability (November/December mating group)
	Bluetongue B (6 month-old lambs)
	Market lambs that are ready (6 month-old lambs)
Nov.	6 week mating period (November/December mating group)
	Blue tongue C, Rift Valley fever, Wesselsbron disease (7 month-old lambs)
	Market lambs that are ready (7 month-old lambs)
Dec.	Avoid working with mated ewes unnecessarily (November/December mating group)
	Market lambs that are ready (8 month-old lambs)

Area : **North Western Karoo**
Type of Grazing : **Veld**
Type of Livestock : **Dorper sheep**
Author : **Carnarvon Experimental Station**
First Mating month : **4** Second Mating month: **10**
First Marketing age: **9** Second Marketing age: **8**

Month	Activity
Jan.	Young ewes Blue udder 1 (March/April lambing group)
	Rev 1 ram lambs to be kept as sires (4 month-old lambs)

	Wean lambs, dose broad spectrum incl. tape worm (4 month-old lambs)
	Market remainder of lambs (10 month-old lambs)
Feb.	Dose all sheep broad spectrum
	Ewes + rams Pulpey kidney
	Ewes Enzootic abortion (April/May mating group)
	Suppl. feeding last 4 weeks pregnancy if necessary (March/April lambing group)
	Ewes Blue udder, young ewes Blue udder 2 (March/April lambing group)
	Rift valley fever + Wesselsbron lambs kept for breeding (5 month-old lambs)
March	Rams fertility testing (April/May mating group)
	Finalise ewe flock to be mated (April/May mating group)
	Flushing of mating flock if required (April/May mating group)
	Put in teaser rams (April/May mating group)
	Ewes are lambing (March/April lambing group)
April	6 Week mating period (April/May mating group)
	Market culled / old ewes (April/May mating group)
	Supplementary feeding lactating ewes if necessary (March/April lambing group)
	Docking and castrating of lambs (1 month-old lambs)
	Market all lambs that are ready (7 month-old lambs)
May	Avoid working with mated ewes unnecessarily (April/May mating group)
	Dose lambs tape worm (2 month-old lambs)
	Lambs Pulpey kidney 1 (2 month-old lambs)
	Market all lambs that are ready (8 month-old lambs)
Jun.	Scan ewes for pregnancy 42 days after mating (April/May mating group)
	Lambs Pulpey kidney 2 (3 month-old lambs)
	Wean ram lambs >30 kg (3 month-old lambs)
	Market all lambs that are ready (9 month-old lambs)
Jul.	Young ewes Blue udder 1 (September/October lambing group)
	Rev 1 ram lambs to be kept as sires (4 month-old lambs)
	Wean lambs, dose broad spectrum incl. tape worm (4 month-old lambs)
	Market all lambs that are ready (10 month-old lambs)
Aug.	All sheep (except weaner lambs) Blue tongue
	All sheep (except weaner lambs) Pulpey kidney
	Dose all sheep broad spectrum
	Ewes Enzootic abortion (October/November mating group)
	Suppl. feeding last 4 weeks pregnancy if necessary (September/October lambing group)
	Ewes Blue udder, young ewes Blue udder 2 (September/October lambing group)
	Rift valley fever + Wesselsbron lambs kept for breeding (5 month-old lambs)
	Market remainder of lambs (11 month-old lambs)

Sept.	Rams fertility testing (October/November mating group)
	Finalise ewe flock to be mated (October/November mating group)
	Flushing of mating flock if required (October/November mating group)
	Put in teaser rams (October/November mating group)
	Ewes are lambing (September/October lambing group)
	Market all lambs that are ready (6 month-old lambs)
Oct.	6 Week mating period (October/November mating group)
	Market culled / old ewes (October/November mating group)
	Supplementary feeding lactating ewes if necessary (September/October lambing group)
	Docking and castrating of lambs (1 month-old lambs)
	Market all lambs that are ready (7
	month-old lambs)
Nov.	Avoid working with mated ewes unnecessarily (October/November mating group)
	Dose lambs tape worm (2 month-old lambs)
	Lambs Pulpey kidney 1 (2 month-old lambs)
	Market all lambs that are ready (8 month-old lambs)
Dec.	Blue tongue all sheep not done Blue tongue in August
	Dose all sheep (except suckling lambs) Nasal worm
	Scan ewes for pregnancy 42 days after mating (October/November mating group)
	Lambs Pulpey kidney 2 (3 month-old lambs)
	Wean ram lambs >30 kg (3 month-old lambs)
	Market all lambs that are ready (9 month-old lambs)

www.ingramcontent.com/pod-product-compliance
Lightning Source LLC
Chambersburg PA
CBHW070726220326
41598CB00024BA/3317

* 9 7 8 0 6 2 0 4 7 8 9 6 0 *